SHOOTING STAR

★★★

China's Military Machine in the 21st Century

SHOOTING STAR

★★

China's Military Machine in the 21st Century

A Report by the Centre for Analysis of Strategies and Technologies (CAST), Moscow

Mikhail Barabanov, Vasiliy Kashin & Konstantin Makienko

east view press

East View Press
Minneapolis, USA

SHOOTING STAR: China's Military Machine in the 21st Century

A Report by the Centre for Analysis of Strategies and Technologies (CAST), Moscow
Mikhail Barabanov, Vasiliy Kashin & Konstantin Makienko

Translated from the Russian by Ivan Khokhotva
Cover Design by Peter Hill Design
Cover Photo: Helicopters fly past Chinese Jiangwei II class naval frigate *Luoyang* in Qingdao, Shandong Province on April 23, 2009. REUTERS/Guang Niu/Pool

Library of Congress Cataloging-in-Publication Data

Barabanov, Mikhail.
   [Oboronnaia promyshlennost' i torgovlia vooruzheniiami KNR. English]
   Shooting star : China's military machine in the 21st century : a report by the Centre for Analysis of Strategies and Technologies (CAST), Moscow / Mikhail Barabanov, Vasiliy Kashin & Konstantin Makienko ; [translated from the Russian by Ivan Khokhotva]. -- 1st ed.
     p. cm.
   Includes index.
   ISBN 978-1-879944-09-1 (pbk.) -- ISBN 1-879944-09-X (pbk.) 1. China--Defenses. 2. China--Armed Forces. 3. Defense industries--China. 4. Arms transfers--China. 5. China--Military policy. 6. China--Foreign economic relations. I. Kashin, Vasiliy. II. Makienko, Konstantin. III. TSentr analiza strategii i tekhnologii (Moscow, Russia) IV. Title.
   UA835.B34 2012
   338.4'735500951--dc23
                 2012009622

Published by East View Press,
an imprint of East View Information Services, Inc.
Minneapolis, USA

Printed in the United States of America

First Edition, 2012
1 3 5 7 9 10 8 6 4 2

Contents

★★★★★★★★★★★★★★★★

Publisher's Foreword

The rise of China is perhaps the single most compelling story of our time, and is having a profound effect on the politics and economics of countries throughout the world. Just as China's political and economic clout has risen, so too has the status of its defense industry and presence on the global arms market. In fact, it's hard to ignore the progress made by the Chinese defense industry in just the last 10 years. From the launch of sea trials for its first aircraft carrier to its expanded offerings on the international arms market, China has made it clear to the world that it is committed to improving and expanding the capabilities of its defense industry.

As such, it is more important than ever to have a firm understanding of the capabilities of China's defense industry—its organization, strengths and weaknesses. The present work, *Shooting Star: China's Military Machine in the 21st Century*, provides just that. *Shooting Star* presents the reader with an objective study of the developments and progress made by China's defense industry over the past 20 years, including an in-depth analysis of China's arms imports and exports.

Shooting Star is the English-language translation of a publication by the Centre for Analysis of Strategies and Technologies (CAST), a Moscow-based think tank specializing in the study of Russia's defense industry, arms procurement and arms trade. We at East View see tremendous value in bringing Russian analysis on this topic to the English-speaking world—as a long-time exporter of arms and various defense technologies to China, Russia is in a unique position to evaluate the changes and developments that have taken place within the Chinese defense industry over the last few decades. As China's defense industry has evolved, Russia has faced shrinking orders from one of its top customers and new competition in its traditional export markets. The authors at CAST do not see this trend stopping anytime soon—there is little doubt that China will become one of the world's leading arms exporters within just a matter of years. And while it is difficult to predict just how far China's star will ultimately rise, what is clear is that China's continued focus on and investment in its defense industry has tremendous implications for the US, Russia, Europe and countries the world over.

East View was quite pleased to have the opportunity to work with CAST. As the publisher of two magazines and numerous books on the subject of defense, CAST is a well-respected voice in the global conversation on defense and arms trade issues. CAST is frequently called upon to provide analysis and consultations for Russian government agencies, defense industry companies and investment firms, and is often quoted in prominent Russian newspapers and other mass media.

I would like to express our appreciation to the numerous individuals who contributed to this book, starting with our partners from CAST: Director Ruslan Pukhov and Deputy Directors Konstantin Makienko and Aleksei Pokolyavin; authors Mikhail Barabanov and Vasiliy Kashin; and translator Ivan Khokhotva. On the East View team, I would like to thank Kent D. Lee, Dima Frangulov and Laurence Bogoslaw for their hard work and support in producing this book.

Ana K. Niedermaier
Director of Publishing
East View Press
April 9, 2012

Authors' Foreword

Over the last decade, China's military-industrial complex has made a qualitative leap forward in building up its technological and economic potential. Since the late 1990s, China has significantly upgraded all aspects of its military production, and Chinese military-industrial corporations such as AVIC, CNGC and CSGC are now among the 500 largest companies in the world.

Such rapid advancement by the Chinese has fueled a need for objective analysis of China's defense industry. In *Shooting Star: China's Military Machine in the 21st Century*, this analysis comes from Moscow's Centre for Analysis of Strategies and Technologies (CAST).

The growth of China's defense industry has great economic, military and political ramifications for Russia and the world. Russia is concerned about the reduction of its arms supplies to China, China's copying of Russian technology, as well as growing competition between Russian and Chinese weapons in third country markets. In addition, despite the friendly nature of present-day Russian-Chinese relations, Russia cannot completely ignore the long-term implications of China's growing military might on its own military security.

For the US, China's strengthening and rearming has become a key factor in military planning. The gradual revision of US military priorities as a result of Chinese growth culminated in January 2012 with the adoption of a new military development doctrine that attaches great importance to expanding its military presence in the Asia-Pacific region.

The possibility of procuring a wider range of more sophisticated weapons from China in addition to Western countries and Russia (which is often vulnerable to Western pressure) is an important factor that could influence the policies of developing countries in Asia, Africa and Latin America. In the future, the role of that factor will continue to grow in relation to the growth of China's industrial potential and foreign policy ambitions.

Shooting Star: China's Military Machine in the 21st Century provides readers with an objective analysis of major trends in the development of China's defense industry and the growth of its export potential in the light of political and macroeconomic factors. We believe that China's defense industry and arms trade are on an upward trajectory, and are confident that in the foreseeable future, the Chinese military-industrial complex will only grow in significance as China develops a new, more active and seemingly inevitable aggressive foreign policy.

Introduction

China's defense industry capability has been growing in leaps and bounds over the past decade. Until the turn of the new century, Chinese defense technology was nothing short of primitive. But the progress it has made since then, along with the entire Chinese economy, is astounding. The product range offered by China's defense contractors bears little resemblance to their archaic offerings of 10 years ago. Indeed, in segments such as aerospace, Chinese manufacturers have leapfrogged two generations of technology. After massive retooling programs, they now boast some of the most advanced manufacturing equipment in the world.

As a result, all the preconditions are now in place for China to transition from a large arms importer into one of the world's leading exporters. The party that will suffer the most from such a transformation is Russia. It's bad enough that it is losing (or may have already lost) its biggest defense customer. Worse, it is also gaining a powerful and dangerous competitor on the global defense market. Both countries work in the same geographic and product segments of that market, namely developing countries with multi-directional—or anti-Western—foreign policies. And both offer moderately advanced weaponry in the medium price range.

The rise of China's defense industry has been accompanied by the overall rise of China's economic and political clout. In the past few years, China has become an important player in several markets that Russia has traditionally regarded as its own turf. The growth of Beijing's political influence is lagging behind the country's economic growth—but it is quite impressive nonetheless. Sooner or later, we are going to see the emergence of a group of developing countries whose economic and political course will follow China's lead. When that happens, Russian arms exporters will find themselves in an even trickier situation, regardless of the quality of their wares.

This book consists of two parts. In the first part, we are going to look at the current state of China's defense industry and its key distinguishing features. We will analyze what a typical Chinese defense company looks like these days. We will parse the structure of government control of the Chinese defense industry, study its institutional landscape, and use the shipbuilding and the aerospace engine sectors for two separate case studies. The choice of these particular sectors as case studies is not incidental. Shipbuilding is the area where Communist China's progress over the past few years has been truly phenomenal. The country has become number one in the world in terms of the civilian shipping tonnage it launches each year. The technology of the military ships on the drawing board or already under construction has also improved dramatically. Nevertheless, the Chinese are still lagging behind the West and even Russia in such areas as naval propulsion, shipborne weapons (especially air defenses) and electronics. In the second case study, which focuses on aerospace engines, China's achievements have been much less spectacular. The technologies and science involved here are far more complex than those required for building ship hulls. Nevertheless, obvious progress has been made in this sector as well. While the country is nowhere near challenging the dominance of the Anglo-Saxon Big Three (GE, Rolls-Royce and Pratt & Whitney), Chinese aerospace design and engineering capability continue to improve, and sooner or later the country will at least achieve relative technological independence in engine-making.

The second part of the book will focus on China's position on the global defense market since the fall of the Soviet Union, i.e., after 1992, with a short summary of the situation in earlier years. We will analyze China's arms imports (mainly from Russia, of course), as well as its growing status as a major defense exporter. The main finding of this section is that Chinese arms imports have fallen sharply since 2005, and are now limited mostly to aerospace and ship engines, transport and attack helicopters, and airborne weapons. Until recently, China had also been taking significant deliveries of air defense weaponry, though most of those deliveries were booked back in 2004-2006. Meanwhile, Chinese defense exports have more than doubled from $800 million a year to $1.8 billion. The main conclusion of this second section and of the book as a whole is unambiguous: Within the next few years, China will not only regain its status as a net exporter of weapons (indeed, it may have done so already), but will actually become one of the world's leading arms exporters.

Chapter 1
China's Defense Industry

1.1 Overview

Almost all of China's defense industry assets are owned by 10 large state-owned corporations. Between them, these companies represent all defense industry segments and conduct nearly all military research and development.

China's 10 key defense corporations

NO.	ENGLISH NAME *	SEGMENT
1	China Aviation Industry Corporation (AVIC)	Aerospace
2	China North Industries Group Corporation (CNGC)	Ground troops weaponry
3	China South Industries Group Corporation (CSGC)	
4	China Aerospace Science & Technology Corporation (CASC)	Missiles and space technology
5	China Aerospace Science & Industry Corporation (CASIC)	
6	China State Shipbuilding Corporation (CSIC)	Shipbuilding
7	China Shipbuilding Industry Corporation (CSIC)	
8	China Electronics Technology Company (CETC)	Electronics
9	China National Nuclear Corporation (CNNC)	Nuclear industry**
10	China Nuclear Engineering & Construction Corporation (CNECC)	

The English company names cited here are used by the Chinese companies themselves in their translated official documents. These names do not always represent verbatim translations of the original Chinese company names.
**This study does not focus on China's nuclear corporations, because they are not involved in arms exports.*

The highest-level decisions concerning the Chinese defense industry are made by the State Council and the Central Military Council (CMC). Decisions to launch large defense industry programs, as well as rules and regulations for the entire industry, are usually issued in the form of joint resolutions by the State Council and the CMC. Every large arms export contract also requires the approval of both councils.

The State Council also oversees agencies in charge of individual economic and technological aspects of Chinese defense industry policies. These include the State-owned Assets Supervision and Administration Commission (SASAC) and the State Administration for Science, Technology and Industry for National Defense (SASTIND). The CMC structure includes the General Armament Department (GAD), which draws up military hardware requirements and specifications, issues procurement orders to the defense industry, and evaluates and tests the weaponry being procured.

All 10 Chinese defense corporations share a set of distinguishing features that define the overall shape of China's defense industry. Each of the 10 is a large conglomerate, employing tens of thousands or even hundreds of thousands of people. For example, as of first quarter 2010, the CNGC employed some 285,000 people.[1] Each corporation has a complex internal structure. CNGC and China Aerospace Science & Technology Corporation have more than 100 subsidiaries each; all of them are directly subordinate to the head company. All Chinese defense corporations own huge amounts of non-core assets. They tend to derive most of their revenues—and in most cases the greater part of their profits as well—from the civilian side of their business. They also invest heavily in areas that have nothing whatsoever to do with their core business, including tourism, luxury goods, and trade in petrochemicals.

Although civilian contracts generate most of their revenues, it is military contracts that earn the Chinese defense industry corporations generous state support. The management performance of these state-owned giants is often judged by the government based on defense contracts. As one CNGC corporate slogan puts it, "Military contracts give us status, civilian contracts give us business."[2]

Some of the corporations have their own foreign trading divisions, such as the China North Industries Corporation (NORINCO Corp., owned by

CNGC) or AVIC International (owned by AVIC). These groups are licensed to sell weapons to foreign customers. Other defense corporations rely on the services of specialized intermediaries, such as Poly Technologies Co. As a rule, these intermediaries long ago turned into large and diversified businesses in their own right. Apart from defense contracts, they also make money from engineering projects abroad, direct investment, and in some cases even mining and trade in minerals.

In recent years, the government has launched an internal restructuring of the defense industry corporations; many of their assets have been merged into subsidiary holding companies. Each corporation now has from six to 30 of these holding subsidiaries, some of which have conducted IPOs on the stock markets of mainland China and Hong Kong. As a peculiarity of the Chinese stock market system, companies can have several types of shares, some of which can be owned by foreigners, and some only by Chinese buyers. This system ensures that foreign investors cannot acquire a controlling stake in any Chinese defense contractor.

The Chinese government encourages private sector involvement in defense contracts; privately owned companies are often used as subcontractors by the state-owned giants. In October 2008, the Ministry of Industry and Information Technology released its "Guidelines on the participation of non-state-owned companies in the development of the defense industry."[3] The document states that "private capital should be encouraged and enticed to participate in the development of the defense industry." It also says that private companies should be encouraged to bid for defense contracts. Nevertheless, the private sector will probably continue to play second fiddle to state-owned corporations in the Chinese defense industry.

1.2 Government Oversight

State-owned Assets Supervision and Administration Commission

The large state-owned companies directly accountable to the government (including the defense industry corporations) are supervised by the State-owned Assets Supervision and Administration Commission (SASAC). The commission can directly intervene in the corporate decision-making process and take part in appraising the corporations' senior managers. Based on State Council resolutions, SASAC can issue binding instructions on a broad range of issues, including financial reporting and personnel policy. The commission also has the power to determine which sectors should be treated as a priority for investment, and which are off limits.

For example, in March 2010 SASAC banned companies whose core business is not tourism or property development from investing in hotels.[4] It also declared property development off limits to all companies whose core business lies elsewhere.[5] The decisions were part of the government's broader economic regulation; specifically, Beijing was trying to cool down the overheated property market.

SASAC also plays the central role in mergers and reorganizations of state-owned companies. There has been a lot of activity in this area in recent years as the government pursued the strategy of creating "national champions." In the Chinese defense industry the last big merger was completed in 2008, when the AVIC aerospace corporation was set up. It appears that the government has no immediate plans for any further restructuring in the defense sector.

State Administration for Science, Technology and Industry for National Defense

The State Administration for Science, Technology and Industry for National Defense (SASTIND) is in charge of setting the overall development strategy for the Chinese defense industry. The agency was set up as part of an administrative reform launched in March 2008. Its predecessor, the Commission of Science, Technology and Industry for National Defense (COSTIND), ceased its existence as an independent ministry-level agency and was subsumed by the Ministry of Industry and Information Technology.

SASTIND's main function is coordinating the work of industrial and research centers on military R&D and on manufacturing orders placed by the People's Liberation Army's General Armament Department (GAD). SASTIND awards contracts and subcontracts to individual companies, in some cases using competitive mechanisms to determine the winner. The agency is also in charge of overall planning for the defense sector. It draws up regulations for the industry, oversees quality control and industrial safety, monitors accounting and compliance, and manages international relationships. Another important part of the agency's purview is the training of engineering specialists for the industry. Until 2008, several leading engineering schools were directly subordinated to COSTIND. Now they are part of the Ministry of Industry and Information Technology, but SASTIND has retained an important role in running them.[6]

Apart from SASTIND, the Ministry of Industry and Information Technology has another department involved in supervising the defense industry — the Department for the Promotion of Integration of Military and Civilian Programs, which oversees efforts to make use of military technology advances in the civilian sector and to harmonize military and civilian technical standards.[7]

General Armament Department

The PLA's General Armament Department (GAD) was set up in 1998; up until then it was part of COSTIND. Before 1998, COSTIND's remit with regard to the defense industry included two conflicting functions. On the one hand, the agency (together with the General Staff and the PLA's General Logistics Department) drew up specifications and requirements for the weaponry being procured by the PLA, and placed orders with individual companies. On the other hand, it oversaw deliveries on the contracts it had itself awarded. Such a conflict of interest led to inefficient use of budget funds, corruption, delays and missed deadlines on arms programs.

The 1998 reform turned the COSTIND departments in charge of drawing up requirements for new weaponry and placing the actual orders into a separate agency, the General Armament Department, subordinated directly to the Central Military Council. The GAD decides what kind of weapons the army needs, draws up specific requirements, and analyzes key global science and technology trends. The department also determines the key areas on which military R&D projects should focus, and evaluates the quality of the weaponry delivered by the industry. To that end, GAD has been provided with its own testing and research facilities. It has also subsumed parts of the General Staff and the General Logistics Department that oversaw R&D policy and military technology policy.[8]

In the relations between the armed forces and the defense industry, GAD represents the interests of the army, and SASTIND those of the industry. Inevitably, the two departments are constantly at each other's throats.[9] Changes in their purview introduced in recent years suggest that the GAD is winning the struggle for preeminence.[10]

1.3 Institutional Landscape

Aviation Industry Corporation of China

The Aviation Industry Corporation of China (AVIC) was formed in 2008 through a merger of China Aviation Industry Corporation I (AVIC I) and China Aviation Industry Corporation II (AVIC II). At the time of its incorporation, the company's authorized capital stood at 64 billion RMB ($9.4 billion). It employed 400,000 people and owned assets worth 290 billion RMB ($42.5 billion).[11]

AVIC has brought under a single corporate roof nearly all of China's civilian aerospace assets, including the makers of military and civilian aircraft, helicopters, engines, avionics, etc. AVIC divisions also supply various types of airborne weaponry, including air-to-air missiles and some types of anti-ship missiles.[12]

The only large Chinese aircraft maker not incorporated into AVIC is the Commercial Aircraft Corporation of China (COMAC), which was also set up in late 2008. COMAC is an independent corporation that reports directly to the government and is run by SASAC. The company was set up specifically to develop China's first long-haul jet airliner, the C919. The project is heavily reliant on foreign suppliers and subcontractors. It appears that COMAC needs to be independent from AVIC so as to make it easier to import foreign components and technologies for the C919, as well as to thwart the lobbying of Chinese component suppliers. It terms of its size, COMAC comes nowhere near AVIC; its payroll is only about 6,000 employees.[13]

The civilian side of AVIC's business, including its non-aerospace assets, generates more revenue for the company than defense or aerospace contracts. In fact, most of the company's sales in the civilian segment are generated by its automotive division. In 2007, auto manufacturing accounted for 67.8% of the revenues of AVIC II, one of AVIC's two corporate forebears.[14] Another big cash earner is industrial equipment and electronics.

The pace of reforms at AVIC has been more rapid than at any other Chinese defense industry giant. Beijing views combat aviation and the aerospace sector in general as one of the main locomotives of future high-tech based economic growth. AVIC's assets are spread among several subsidiary holding companies, each specializing in its own product or technology segment to avoid internal competition. These subsidiaries enjoy a fair degree of autonomy, and each has its own long-term development strategy. The formal incorporation of all AVIC divisions has been completed, but the actual redistribution of assets between them will continue well into 2012.

AVIC subsidiaries

NO.	HOLDING COMPANY	SPECIALIZATION
Manufacturing divisions		
1	AVIC Aircraft	Aircraft (mainly military)
2	AVIC General Aircraft Company	General-purpose aircraft
3	AVIC Helicopter Company (Avicopter)	Helicopters
4	AVIC Aero-Engine Company	Aircraft engines (civilian and military)
5	AVIC Commercial Aircraft Engine Company	Aircraft engines (civilian)
6	AVIC Systems	Aircraft systems and components
Administration and management		
7	AVIC Defense Industry Division	中航工业防务事业部 Supervision of all defense projects
8	AVIC Assets Management Division	中航工业资产管理事业部
9	AVIC International	Supervision of all non-core projects 中航工业技术国际控股有限公司 Overseas operations
10	Avichina Industry & Technology Company Ltd.	中国航空科技工业股份有限公司 Investment in high-tech projects

NO.	HOLDING COMPANY	SPECIALIZATION
R&D and testing		
11	AVIC Fundamental Research Technology Institute	中航工业技术基础研究院 Fundamental research
12	AVIC Economic Research Institute	中航工业经济技术研究院 Economic research
13	AVIC Flight Test Establishment	中国飞行试验研究院 Testing facilities
Other assets		
14	AVIC Automobile	中航工业汽车公司 Auto manufacturing
15	Joy Air Ltd.	幸福航空有限责任公司
16	AVIC Investment	Airline
17	AVIC Construction Projects Company	中航工业投资公司 Property investment 中航工业建设工程公司 Construction

Source: AVIC corporate website (www.avic.com.cn)

AVIC Aircraft controls all the main Chinese makers of civilian and combat aircraft, including Shengyang Aircraft, Chengdu Aircraft, Xian Aircraft, and Shaanxi Aircraft. The division will also be in charge of all the main military R&D projects in the aerospace sector. It is also involved in civilian aircraft projects led by COMAC, but its main line of business will always be defense technology.

AVIC General Aircraft Company will specialize in making light and trainer aircraft. The division now controls the Shijiazhuang Aircraft Industry Company, a manufacturer of light aircraft, and the Guizhou Aircraft Industry Group (GAIC), which makes combat trainers. The division is also building a new light aircraft production facility in Zhuhai. Together with GAIC, AVIC General Aircraft Company also controls the production of Chinese unmanned aerial vehicles.

Avicopter is set to unite China's main helicopter industry assets, including:

- Helicopter Research & Development Institute No. 602 in Jingdezhen;
- Hafei Aviation Industry Co. (Harbin Aircraft Industry Corporation), maker of the Z-9 helicopter;
- Changhe Aircraft Industry Corporation, maker of the Z-8 and Z-11, as

well as the future WZ-10 attack helicopter;

- Baoding rotor plant, China's only manufacturer of rotors for helicopters and for piston-engine and turboprop aircraft.

Avicopter is also building a new manufacturing center in Tianjin.

AVIC Aero-Engine Company will make engines for civilian and combat aircraft. *AVIC Commercial Aircraft Engine Company* will specialize in civilian engines. The main difference between the two is that, like COMAC, AVIC Commercial Engine will make use of foreign components, technologies and expertise. The company plans to hire up to 2,000 foreign specialists in the next few years.[15]

AVIC Systems will specialize in fuel and hydraulic systems and other aerospace components. It is not clear at this time what assets the division will control.

AVIC Defense Industry Division is an administrative outfit in charge of coordinating all AVIC defense programs. Judging from media reports, it appears that the division will not actually control any defense industry assets of its own. *AVIC Assets Management Division* will manage non-core assets, which any state-owned Chinese corporation has in spades (hotels, resorts, education facilities, etc.).

AVIC International was formed in June 2009 through the merger of CATIC,[16] a well-known Chinese foreign trading group that specialized in aerospace products, with China Aviation Industry Supply & Marketing and Beijing Raise Science. The CATIC brand is used by many of AVIC's former subsidiaries.

Avichina Industry & Technology Company manages investment in aerospace and related high-tech sectors. It owns controlling stakes in several aerospace equipment manufacturers, including helicopter plants in Harbin and Jingdezhen that make the Changhe helicopter. The company is gradually shifting its focus to high-tech aviation components, primarily electronics.

R&D and testing centers have the same status within AVIC as manufacturing divisions, but they play an auxiliary role and do not seem to be slated for any further structural reforms.

Finally, *AVIC Automobile* controls the corporation's huge assets in the automotive industry. Almost every large Chinese aerospace company owned some automotive assets prior to the latest reform.

Senior AVIC executives see Avicopter and AVIC General Aircraft Company as the main engine of growth for Chinese aerospace exports. The current target is for the latter to win one-third of the world market for light aircraft by 2025,[17] and for the former to seize 15% of the world market for helicopters in the next 15-20 years.[18]

One of the central planks of AVIC corporate strategy is flotation, first of its subsidiaries and eventually of the whole corporation. The process should be largely completed by 2013, when, according to AVIC executives, some 80% of the corporation's subsidiaries will have their shares traded on the stock market. Once that happens, the corporation itself will start preparing for an IPO.[19]

China North Industries Group Corporation and China South Industries Group Corporation

China's Northern and Southern industry groups, CNGC and CSGC, were formed in 1999 by splitting the China Ordnance Industry Group, the COIG, in two. The northern group, CNGC, inherited almost all of the military assets.[20] The southern CSGC has since become one of the world's Top 500 corporations. Its annual revenues have topped 200 billion RMB ($30 billion).[21] CSGC is one of China's largest manufacturers of various machinery and cars, but the defense side of its business is relatively small. The company is known to make small arms and explosives.

CNGC, meanwhile, has become the PLA's sole supplier of most types of ground weaponry. Its foreign trade operations have been spun off into a separate company, the China North Industries Corporation (NORINCO Corp., not to be confused with NORINCO Group, which is another name for CNGC itself). NORINCO Corp. is 50% owned by CNGC. The other 50% is owned by CSGC, which also uses NORINCO Corp. for civilian and defense export operations.

The CNGC's defense product range includes:

- tanks, wheeled and tracked armored personnel carriers (APCs), and special-purpose vehicles using their chassis
- cannon and rocket artillery systems
- small arms
- ammunition for small arms, cannon artillery, multiple launch rocket (MLR) systems and tanks, including guided ammunition
- aviation bombs
- depth charges
- fire control systems
- night vision imagers
- communication equipment
- air defense systems (anti-aircraft (AA) artillery, PL-9 surface-to-air missile (SAM) systems)
- explosives

The CNGC's civilian output includes:

- heavy trucks and buses
- motorcycles
- heavy wheeled and tracked machinery such as bulldozers and dump trucks
- chemicals
- railway equipment and machinery
- petrochemicals

The proportion of CNGC's civilian to defense revenues is 70:30, but defense contracts generate half of the company's net profit.[22] Figures are also available for one of the corporation's key divisions, the Inner Mongolia Machine-Building Company (based in the city of Baotou),[23] also known as Plant No. 617. The company is probably the world's largest manufacturer of main battle tanks.[24] In 2007, it reported that "six years previously the proportion of our military to civilian output was 8:2, but now it has reversed," mainly thanks to growing sales of heavy trucks.[25]

CNGC is a huge corporation; it employs 285,000 people and has 111 subsidiaries. In 2009, its net profit rose to 5.2 billion RMB ($750 million) on operational revenues of 161.8 billion RMB ($24.4 billion) and the combined value of its assets stood at 200 billion RMB ($30 billion).[26] Based on 2009 operational indicators, the corporation said that it had overcome the consequences of the world economic crisis.[27] The diagrams below show the structure of CNGC's workforce.

Structure of CNGC workforce by age and education, as of late 2009 (%)

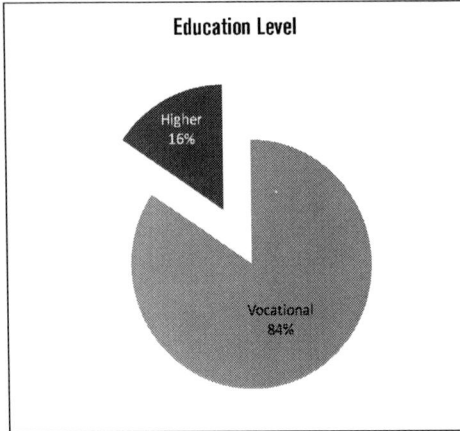

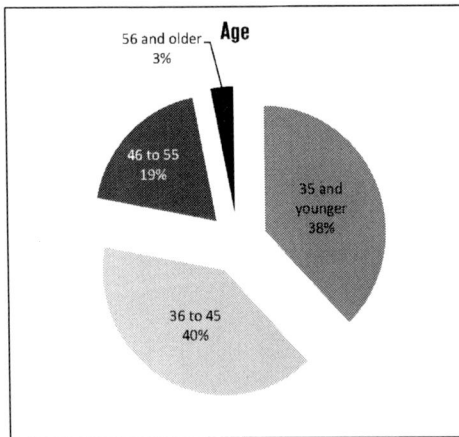

Education Level

Higher
16%

Vocational
84%

Age

56 and older
3%

46 to 55
19%

35 and younger
38%

36 to 45
40%

Source: based on China Lianhe Credit Ratings report of April 30, 2010 on the decision to assign a corporate debt rating to CNGC bonds (http://bond.hexun.com/upload/zhongguobingqi.pdf).

Another important part of CNGC's business is oil and petrochemicals. The corporation owns CNGC Huajin Chemicals.[28] For its part, Huajin Chemicals controls Zhenhua Oil, a company set up in 2003 to pursue oil exploration projects abroad.[29] Zhenhua is a large oil trader and holds a government license for oil imports. It is now building its own refinery with an annual capacity of 5 million tonnes of oil in Liaoning Province. It is also pursuing oil drilling projects in Pakistan, Syria, Kazakhstan and Iraq. Its revenues from selling oil topped $3 billion in 2009.[30] As an oil trader, Zhenhua works in Russia, Kazakhstan, Singapore, South Korea, Oman and countries in West Africa.[31]

The government wants to simplify the CNGC's unwieldy corporate structure, with its 111 subsidiaries, to avoid duplication and cross purposes. It will use the reform of the AVIC aerospace corporation as a template.[32] Over the next three years, CNGC's numerous subsidiaries will be merged into 30 specialized holding companies. Ten of them will be large enough to win a significant share of the relevant segments of Chinese and foreign markets. The holding companies will focus on CNGC's three core product categories: weapons, heavy civilian machinery (dump trucks, heavy engineering machinery, etc.), and equipment for the chemical industry.[33]

China Aerospace Corporation and China Aerospace Science and Industry Corporation

China Aerospace Corporation (CASC) and China Aerospace Science and Industry Corporation (CASIC) were formed in 1999 by splitting the former China Aerospace Corporation (old CASC) in two. Unlike AVIC and CNGC, which have become near-monopolies in their respective segments of the defense market (i.e., combat aircraft and ground weaponry), CASC and CASIC have a partially overlapping product range and compete with each other in many areas. SASTIND is trying to keep their rivalry in check by making the two work together on joint projects. Both companies can export weaponry via the China Precision Machinery Import & Export Corporation (CPMIEC), an intermediary that they own 50-50.

CASC designs and manufactures rocket launchers, ballistic missiles, satellites and the Shenzhou manned spacecraft. It also makes artillery systems (including the A100, WS-1 and WS-2 long-range MLR systems) and air defense systems. It employs 120,000 people. Its key divisions are:

China Academy of Space Launch Technology (CALT), which designs and manufactures liquid and solid-fuel ballistic missiles (including China's primary ICBM, the DF-5), and launch vehicles (CZ-1, CZ-2 and CZ-3). CALT also runs eight R&D centers, an experimental manufacturing facility and two manufacturing plants.[34]

Shanghai Academy of Space Flight Technology (SAST) is similar to CALT in terms of its size and scope. It designs and builds ballistic missiles and launch vehicles (CZ-4, participation in the CZ-2 and CZ-3 programs) and satellites (here it competes with CAST). It also makes the FN-6 man-portable SAM systems and the LY-60 SAMs, which compete head to head with systems made by CASIC.

Academy of Aerospace Solid Propulsion Technology (AASPT), a.k.a. the 4th Academy, is China's primary manufacturer of solid-fuel engines for ballistic missiles and guided SAM missiles. The company employs about 10,000 people; it has six research centers and four manufacturing plants.

China Academy of Space Technology (CAST) makes satellites and the Shenzhou manned spacecraft. It has 10 research centers, a manufacturing plant and a testing facility. It employs 1,700 highly qualified engineers and specialists.[35]

Academy of Liquid Propulsion Technology (AALPT) is China's only manufacturer of liquid-fuel engines for launch vehicles and ballistic missiles.[36]

China Academy of Space Electronics Technology (CASET) makes space components. Several years ago, its employees numbered more than 5,000.[37]

Sichuan Aerospace Industry Corporation (SCAIC) is China's main manufacturer of heavy MLR systems. It is based in the city of Chengdu.

CASIC is barely involved in the manufacture of launch vehicles, but it competes with CASC in the segment of solid-fuel ballistic missiles. Its products include the DF-11 tactical missile, a rival of the DF-15. The company also makes China's primary medium-range ballistic missile, the DF-21.

CASIC has much stronger positions than CASC in SAM systems. In fact, it is China's leading manufacturer of this type of weaponry. It makes the HQ-9, KS-1 and HQ-7 systems, as well as several types of man-portable SAMs marketed under the QuinWey (QW) brand. It also makes the DH-10 medium-range cruise missiles, the YJ-62 anti-ship missiles, and their airborne versions.

CASIC employs 110,000 people, 40% of whom have a higher education.[38] Its key divisions are:

China Changfeng Electromechanical Technology Design Institute (CCETDI), the designer and manufacturer of SAM systems such as the HQ-9 and KS-1, as well as homing systems, optical-electronic equipment and other electronics.

China Haiying Electromechanical Technology Academy (CHETA), the designer and manufacturer of cruise and anti-ship missiles.

China Hexi Chemical and Machinery Company (CHCMC), the maker of solid fuel for rocket engines used in ballistic missiles and maneuvering engines of spacecraft.

China Sanjiang Space Industry Group Company (CSSG) is a large missile corporation with headquarters in Wuhan. It makes solid-fuel ballistic missiles and other weaponry, such as self-propelled transporter-launchers for cruise and ballistic missiles, which are designed and manufactured in cooperation with Belarus's Minsk Wheeled Chassis Plant. The company employs 15,000 people, including more than 6,000 engineers.[39]

CASC and CASIC also make various civilian equipment, including telecommunications and information systems, industrial machinery, refrigeration equipment, mining machinery, medical devices, measuring instruments, and industrial software. The two companies offer their services on the market for civilian space launches via a joint intermediary, the China Great Wall Industry Corporation (CGWIC). From 1985-2008, the company conducted 29 commercial space launches and put 35 satellites into orbit.[40]

China Shipbuilding Industry Corporation and China State Shipbuilding Corporation

China Shipbuilding Industry Corporation (CSIC) and China State Shipbuilding Corporation (CSSC) together control up to 70% of China's shipbuilding assets. They were created as part of a 1999 reform, in which the old shipbuilding monopoly (also called CSSC) was split into two operations.

The company was divided along geographic lines. CSIC inherited assets mainly in northeastern and central China, and the new CSSC in the east and south of the country. The China Shipbuilding Trading Company (CSTC), a division of CSSC, is a specialized Chinese exporter of military ships.

CSIC employs 140,000 people; it has 28 research centers and 46 production facilities, including seven shipbuilding plants.[41] In 2009, it made $1.1 billion of net profit on $18.1 billion of revenue.[42] CSSC employs 100,000 people.[43] In 2009, it made $366 million on $3.4 billion of revenue.[44]

Both firms earn a lot more revenue from the civilian side of their business than from defense contracts. CSIC owns the Liaoning Shipbuilding Group, which controls the Bohai Shipyard in the city of Huludao, China's only nuclear submarine manufacturer. It also owns the old and the new shipbuilding plants in Dalian, which build fleet destroyers (Type 051B) and tank landing ships (Type 072). The Dalian shipyards are refitting the *Varyag*, a half-finished Soviet heavy aircraft-carrying cruiser that China bought from Ukraine in 2000. CSIC also controls the Qingshan shipyards in Wuhan, a maker of diesel-electric submarines.

CSSC owns the Jiangnan and Hudong-Zhonghua Shipyards in Shanghai, which build Type 054 frigates, Type 052B and 052C fleet destroyers, and China's latest Type 071 helicopter-carrying amphibious transport dock (Kunlunshan class). The company also controls most of China's production of missile boats and patrol boats (Huangpu Shipyard in Guangdong Province and Xijiang Shipyard in Guangxi Zhuang Autonomous Region).

China Electronics Technology Corporation

China Electronics Technology Corporation (CETC) was founded in 2002. It is China's main manufacturer of military automated command-and-control systems, radars (including long-range detection systems), optical-electronic components, and special communication, electronic warfare and electronic reconnaissance systems. CETC also makes a wide range of civilian electronics, including television equipment, medical devices, measuring instruments, control systems, and electronic components.[45]

CETC employs over 80,000 and has 55 subsidiaries,[46] mainly the numbered defense research centers involved in R&D and manufacturing programs.[47]

CETC has gone further than most other Chinese defense corporations in engaging the private sector. It works with such well-known industry players as Huawei Technologies, Great Dragon Telecommunications Equipment Company, Zhongxing Telecommunications Equipment, and Datang Telecom Technology Company.[48] There is a lot of movement of technical and managerial personnel between these commercial companies and the numbered R&D centers run by the CETC.

CETC is a large exporter of civilian IT products; in 2009, its exports reached $1.2 billion.[49] The company does not have its own division licensed by the government to export defense products. Military exports are therefore channeled via China National Electronics Import & Export Corporation (CEIEC), an independent trading company. In recent years, CEIEC has evolved into a large and diversified state corporation. The export of military electronics is only one of its many lines of business, in addition to trade in various civilian products, large engineering projects abroad, and exports of ships and components.[50]

1.4 Case Study: Shipbuilding

Over the past decade, Chinese naval shipbuilding has made rapid progress. The government, empowered by the country's unprecedented economic growth, is aiming to turn China into a leading military and naval power. The pace of rearmament programs in the PLA's Navy has been nothing short of astounding. China is building huge numbers of new-generation warships and boats. They are being equipped with relatively advanced weapons and electronic systems. Until the second half of the previous decade, China had placed large orders with Russia. Now, however, it has achieved almost complete self-sufficiency in building both surface ships and submarines—although it remains dependent on supplies of some systems (missiles, electronics, gas turbines) from Russia and Ukraine.

The phenomenal rise of Chinese naval shipbuilding was made possible by an unprecedented growth in commercial shipbuilding. In 2010, China claimed the crown of the world's largest builder of commercial ships. Over the past decade, a large number of large and well-equipped shipyards have been built in the country; their production capacity is already huge, and more is being added every year.

The two poles of Chinese shipbuilding

The current structure of the Chinese shipbuilding industry reflects the specifics of the country's economic model, as well as China's long-standing tradition of dividing its naval power into the "Northern" and "Southern" fleets or squadrons. (At this time, the "Southern" component of the Chinese Navy includes the Eastern and Southern Fleets.) As a result of that tradition, China has always aimed to create separate and largely self-sufficient shipbuilding hubs to cater to the needs of its separate fleets.

In 1982, as part of the first stage of its economic reforms, China restructured its 6th Machine-Building Department (in charge of the shipbuilding industry) as the China State Shipbuilding Corporation, the CSSC. The Chinese shipbuilding assets were thereby put on a commercial footing under the umbrella of a single commercial giant, with highly qualified quasi-ministerial level management. That strategy was very different from the course later chosen by the former Soviet Union and Russia, where the shipbuilding industry became fragmented into a lot of small and struggling companies.

In July 1999, as part of an overall reorganization of China's defense industry, CSSC was split into two large state-owned corporations. One retained the CSSC name and all shipbuilding assets in the south of the country. The other, the China Shipbuilding Industry Corporation (CSIC), took over the shipbuilding operations in the northern part of China. Naturally, such geographic division is not absolute; CSSC has some assets in the north, and CSIC some in the south.

At this time, the "northern" CSIC corporation, with headquarters in Beijing, controls 97 separate companies with a total payroll of more than 300,000 people, though not all of these assets are in the shipbuilding sector. The corporation's subsidiaries include seven shipbuilding companies in Dalian, Tianjin, Wuhan, Chongqing, Huludao and Kunming; 52 subcontractors and specialized plants; 28 research institutes and 10 laboratories; and 15 high-tech companies in sectors other than shipbuilding. CSIC's overall shipbuilding capacity is estimated at 7 million gross tonnes per year. The corporation's biggest asset is the Dalian Shipbuilding Industry Company (Group), which is the key maker of ships for the Northern Fleet. Another big subsidiary, the Xian Marine Equipment Industry Company, produces various ship components and systems. CSIC also owns four trading companies operating outside China. In order to attract investment from shareholders, in 2008 CSIC set up a publically traded subsidiary.

The China State Shipbuilding Corporation (CSSC), with headquarters in Shanghai, is based in the south. It includes about 50 production companies and nine research centers, including large shipyards in Shanghai and Guangzhou, as well as the country's leading Marine Design & Research Institute of China (MARIC). It also has a foreign trade subsidiary, the China Shipbuilding Trading Company. The overall CSSC payroll is about 150,000 people. The corporation is growing at a very impressive rate, ramping up the output of the existing shipyards and launching new ones. In the past decade, it has built a new shipbuilding and ship repair plant on Chongming Island, near Shanghai, and the world's largest shipyards in Longxue, near Guangzhou, and on Changxing Island, also near Shanghai.

Although it employs fewer people than CSIC, CSSC remains China's largest shipbuilding company; in 2010, it ranked as the world's third largest by sales volume. It has set itself an ambitious target of launching 14 million tonnes of ships per year by 2015, thereby claiming the top place in the world ranking. The corporation has two separate shipbuilding hubs in Shanghai and Guangzhou, serving the Eastern and the Southern fleets, respectively.

Both of China's shipbuilding giants have been implementing serious restructuring programs in recent years. They have been merging their subsidiaries into larger territorial or specialist units, some of which have become publically traded companies, with the CSIC and the CSSC retaining controlling stakes. The bulk of the Chinese shipbuilding companies' business comes from civilian contracts; military shipbuilding plays a secondary role. None of the shipyards are designated as purely military—not even those that were built many years ago with the sole purpose of making warships (such as the Bohai shipyard in Huludao).

Apart from the shipbuilding assets operated by CSIC or CSSC, there are also many smaller shipbuilding and ship repair companies in China. They are owned mostly by large private or semi-private industrial conglomerates, provincial administrations, municipalities or large state-owned shipping companies. But the involvement of these smaller companies in military shipbuilding is negligible. There are also five joint shipbuilding companies with foreign capital, and six ship repair centers operated by the Chinese Navy. According to government statistics, in 2009 there were 1,242 companies in China's shipbuilding sector, employing some 460,000 people.

Civilian shipbuilding

Rapid growth of commercial shipbuilding in China made the country the world's third largest shipbuilding power (after South Korea and Japan) in 1994. In 2009, China overtook Japan in terms of the number of ships built, to become the second in the world. In 2010, it overtook South Korea, claiming the top spot in the world ranking. In the period from 2003 to 2010, the number of seagoing commercial ships with a gross tonnage of 100 and above built in China increased sevenfold, from 203 to 1,402 units per year. Their combined gross tonnage increased almost tenfold, from 3.763 million to 36.239 million. In 2010, China accounted for 37.7% of global launches of new ships in tonnage terms, and for 43% of all new orders. By 2011, China's shipbuilding sector had outstanding orders for 2,611 sea-going freighters, with a combined gross tonnage of 101.844 million.

According to government figures, the overall number of ships built in China in 2008, including small vessels, river ships, fishing vessels and warships, was 2,385, and 861 of these were large seagoing commercial ships. China's share of the world shipbuilding market does not look quite as large as these figures suggest if one measures it in compensated gross tonnage (CGT), i.e., if the raw tonnage is adjusted for the complexity of the ship. That discrepancy highlights the fact that the Chinese shipbuilders still tend to specialize in relatively simple types of ships, although they are gradually catching up with their Japanese and South Korean competitors. To illustrate, as of early 2011, dry bulk carriers accounted for some 53.7% of the Chinese order portfolio measured in CGT; oil tankers accounted for 9.7%.

China's commercial shipbuilding sector is heavily dependent on foreign contracts. Over the past decade, some 70% to 78% of large commercial ships built in the country were destined for export. As of early 2011, the share of export contracts in the overall portfolio was 75.2%.

The military side of the Chinese shipbuilding industry is also growing rapidly, thanks mainly to new programs to build surface ships of the main types, as well as assault landing craft. But China is still lagging behind the leading naval powers in terms of the technologies used in naval weapons systems and electronics. It therefore remains dependent on imports of naval weaponry, components, electronics, and propulsion units, which are sourced mainly from Russia and Ukraine. That dependence is one of the reasons why China has not yet become a major exporter of warships.

Chinese military shipbuilding programs

Over the past several years, China has been ramping up its sea power at an unprecedented pace, launching mass production of new-generation submarines and surface ships. Projects to develop new nuclear-powered submarines, including ballistic missile subs, have entered the stage of practical implementation. New non-nuclear powered submarines are being built at a rate of up to four submarines per year. The Chinese Navy is receiving large numbers of increasingly advanced frigates, assault landing craft and missile boats with wave-piercing hulls. It has also received prototypes of radically new types of helicopter-carrying amphibious dock ships, minesweepers and auxiliary ships. The Chinese defense industry has developed new fleet destroyers equipped with advanced medium- and long-range SAM systems supplied by Russia (Shtil-1 and Rif-M) or developed domestically (HQ-9). Over time, these new systems will enable the Chinese Navy to address the inadequacy of its ships' air defenses, which remains one of its main weaknesses. The medium- and long-range SAM systems being fitted onto fleet destroyers and frigates, as well as the new short-range artillery and SAM systems, will also give the Chinese Navy the capability to defend itself against anti-ship missiles.

Finally, in 2011, the shipyards in Dalian completed the *Varyag*, a half-finished Soviet aircraft carrier that China bought from Ukraine in 2002. The ship will become a training and testing platform for China's aircraft carrying fleet; at some point in the future the country, is expected to start building domestically developed aircraft carriers.

On the whole, China is clearly working to build up its naval power in the near ocean zone. The emphasis is on acquiring a large fleet of submarines, fleet destroyers and frigates in order to defend the country's maritime communications and enable it to impose a naval blockade on Taiwan, if need be. The program to build large numbers of assault landing craft also has Taiwan in its sights. The task of ensuring Chinese naval presence in the far oceans or taking on the US Navy is not a priority for now. The naval maneuvers that China has been undertaking from time to time far from its shores seem little more than training events or PR stunts.

At present, China has built up the world's largest navy in terms of the sheer numbers of ships and boats—but that is thanks mainly to the large numbers of smaller units that have entered service. If only frigates, fleet destroyers and submarines are taken into account, the US Navy is still number one. And in terms of the combined non-nuclear fighting ability of its navy, China is second not only to the US, but to Russia as well. The Russian Navy has a significant advantage in the numbers of nuclear submarines, supersonic anti-ship missiles, and the overall level of technology. In addition, the Chinese Navy's ability to defend itself against submarine and air attacks, as well as mines, remains very limited, given the nature of modern warfare. In terms of technology rather than sheer numbers, China is not expected to catch up with the leading Western naval powers anytime soon.

Main Chinese military shipbuilding companies

China Shipbuilding Industry Corporation (CSIC) subsidiaries

At present, CSIC controls five regional shipbuilding groups:

- Dalian Shipbuilding Industry Company (Group)
- Wuchang Shipbuilding Industry Company
- Chongqing Shipbuilding Industry Company
- Tianjin Xingang Shipbuilding Heavy Industry Company
- Shanhaiguan Shipbuilding Heavy Industry Company
- Beihai Shipbuilding Heavy Industry Company (shipyard currently under construction)

Dalian Shipbuilding Industry Company (Group) (DSIC) was formed in 2005 through a merger of two large shipyards (Dalian Shipyard and Dalian New Shipbuilding Heavy Industries Corporation). The group also owns the Bohai Heavy Industry Company in Huludao (more on this later).

The Dalian Shipyard company originated as Russian Ship Repair Works in 1898 in what was known back then as the town of Dalny. Later on, the company was taken over by the Japanese. From 1945 to 1953, it belonged to the Soviet Union. In the 1950s, it was handed over to the Chinese. The Soviet Union helped China to modernize the shipyards, which were significantly expanded and renamed the Red Flag Shipyards. The company became one of the leaders of the Chinese military shipbuilding industry. In the 1990s, it underwent a radical modernization program using Japanese technology. As for Dalian New Shipbuilding Heavy Industries Corporation, the shipyard was built in the 1980s and 1990s to serve the civilian market. The combined area of the DCIS shipyards in Dalian after the merger is 2.5 square kilometers. The company employs some 16,000 people.

It has two large dry shipbuilding docks. One is 550 meters long and 80 meters wide, with an overhead crane that lifts 600 tonnes. The other is 440 x 96 m, with a 900 t crane. Both can be used to build ships of practically any size or type. There are also two large building berths (306 x 50 m and 250 x 76.4 m), three covered berths for building warships, and two dry docks for ship repairs. Apart from commercial ships, the company also builds oil and gas drilling platforms.

DSIC has a large civilian business; most of the commercial ships it builds are destined for export. In 1996 Dalian Shipyards became the first company in China to build a 300,000 GT ship. The first warship was built there some 50 years ago. In the 1960s, the company launched its first diesel-powered missile-carrying submarine, built to Soviet Project 629 design (Chinese designation Type 031, Golf class). Later on, it launched mass production of Type 051 fleet destroyers (Luda class). It launched a Type 051B (Luhai class) destroyer in 1998, a Type 918 (Wolei class) minelayer, and the first two of the Fuqing class refueling tankers. In the past decade, it built two Type 051C destroyers (Luzhou class, launched in 2006-2007), three Type 072-III large tank landing ships (Yuting II class), and several Yunshu class medium tank landing ships. Nevertheless, there are some signs that the military side of DSIC's business is in decline.

In 2005, Dalian Shipyards started working on the *Varyag*, the half-finished Soviet Project 11436 aircraft-carrying cruiser, which was towed to Dalian from Nikolayev (Ukraine) in 2002. The project has already been completed, and on August 10, 2011, China's first aircraft carrier entered a sea trials program.

Dalian is also home to the Dalian Marine Diesel Works company, which is part of the CSIC corporation. The company builds low-speed diesel ship engines under Western licenses.

Bohai Heavy Industry Company (Huludao), the former Plant No. 431, was built in Huludao using special Soviet designs and Soviet technical assistance. Work on the site began in 1953 and was completed in the early 1960s. From the very beginning, the plant was meant to build warships; indeed, it still remains the only shipyard in China to build nuclear-powered submarines. In the 1970s and 1980s, it built all five of China's Type 091 (Han class) attack nuclear-powered submarines, and the country's only Type 092 (Xia class) nuclear-powered ballistic missile submarine. Over the past decade, the company has launched production of new-generation Chinese nuclear-powered submarines, including the Type 093 (Shang class) attack nuclear-powered submarines and the Type 094 (Jin class) nuclear-powered ballistic missile submarines. Two subs of each type have already entered service with the Chinese Navy, and several more are on the way. The Bohai shipyards also perform medium-grade repairs on Chinese nuclear-powered submarines. But due to the sluggish pace of the Chinese nuclear-powered submarine program, starting in 1994 civilian contracts have made up the larger part of the Bohai Heavy Industry Company's business.

The design and specifications of the Bohai shipyards are similar to those of the Soviet shipyards in Severodvinsk and Komsomolsk-on-Amur. Ships are built in covered berths and launched via transfer launching docks. The main building berth has seven separate shipbuilding lines, cranes that can lift up to 640 t, and a launching dock that can accommodate ships of up to 40,000 GT. There is also an open building dock, 294 m x 50 m in size, equipped with a 480 t crane, capable of accommodating ships of up to 174,000 GT. The company is now building a separate 480 x 107 m dry dock, which will be used to build commercial ships. The company occupies an area of 4.5 sq.km. in Bohai, and employs an estimated 27,000 people.

Wuchang Shipbuilding Industry Company's main asset is the Wuhan Shipbuilding shipyards in the city of Wuhan, on the Yangtze river, far inland from the sea. For a long time, it was China's main producer of non-nuclear submarines. Founded in 1934 to build river ships, it underwent a modernization program in the 1950s and 1960s. In the past decade, the shipyard was expanded; civilian contracts, including orders for small freighters and supply vessels for offshore drilling, now account for the bulk of its business. The shipyard occupies an area of about 1 sq.km. It has seven covered assembly berths, three open berths and one slip. It employs 8,200 people.

The Wuchang Shipbuilding shipyard played a key role in the program to build large numbers of Type 033 (Soviet Project 633, Romeo class) diesel-electric subs; later on, it built all of China's Type 035 (Ming class) submarines. In the 1990s, the shipyard built the first of the new Type 039G (Song class) submarines, which entered service with the Chinese Navy in 1999. From 2001 to 2006, it launched three modified Type 039G submarines and seven Type 039G1 subs, building them at a rate of up to three per year. After 2000, it launched production of new-generation Type 039A and 039B non-nuclear submarines (Yuan class, sometimes designated as Type 041). The first such submarine was launched in 2004 and entered service in 2006. Another three have been delivered to the Chinese Navy since then, and several more are on the way. In 2011, the plant finished a non-nuclear sub of a new type.

Wuchang Shipbuilding also performs medium-grade repairs on non-nuclear subs and builds surface warships. It used to build Type 010 minesweepers (T-43 class, a copy of Soviet Project 254) and Type 072 (Yukan class) large tank landing ships, as well as reconnaissance and auxiliary vessels. In 2003-2004, the shipyard launched four Type 072-III (Yuting II class) tank landing ships. In recent years, it has launched production of the new Type 081 (Woichi class) minesweepers; at least three have already been delivered to the Navy.

Chongqing Shipbuilding Industry Company owns several shipbuilding assets in Chongqing, including a relatively small shipyard. It specializes in ship components and small ships. It also supplies a number of components for submarines to Wuchang Shipbuilding. Previously it built anti-submarine and patrol boats for the Chinese Navy, but no recent information about the military side of the company's business is available.

Shanhaiguan Shipbuilding Heavy Industry Company is a shipyard on the coast of Bohai Bay, specializing in shipbuilding and ship repairs. Founded in 1972, it occupies an area of 2.16 sq.km. It has three dry docks: 340 x 64 m (for ship repairs), 240 x 39 m and 170 x 28 m. It used to build patrol and anti-submarine boats, but it has not had any military contracts since 2000.

Tianjin Xingang Shipbuilding Heavy Industry Company, situated in Tanggu, near Tianjin, on the coast of Bohai Bay, specializes in building and repairing commercial ships. It employs 6,300 people and builds large numbers of medium-sized commercial vessels. It has two dry docks (212 x 27 m and 106 x 16 m), a floating dock and six slips. In 2011, the shipyard launched the *Azmat,* a 500 t missile boat built under a Pakistani contract.

Beihai Shipbuilding Heavy Industry Company's main asset is Qingdao Haixiwan Shipbuilding and Ship-repairing base, a 3.5 sq.km. shipyard now being built near Qingdao. It will include two shipbuilding dry docks (530 x 133 m and 480 x 96 m, each equipped with two 120 t portal cranes); three ship-repair dry docks (360 x 78 m, 325 x 58 m, and 250 x 45 m); and an 80,000 t floating dock. Shipbuilding operations began here in 2010; ship repairs are expected to commence in 2011-2012. Once the shipyard is fully operational by 2015, its annual shipbuilding capacity should reach 4.68 million GT.

The equipment of the existing Qingdao Shipyard, founded back in 1949, will be relocated to the new site. The company specializes in small commercial ships, patrol and anti-submarine boats, small assault landing craft and auxiliary ships.

China Shipbuilding State Corporation (CSSC) subsidiaries

CSSC shipbuilding operations are concentrated in two large hubs, Shanghai and Guangzhou, and a smaller hub in Jiujiang, Jiangxi Province. It has three regional sub-holdings:

- Shanghai Shipbuilding Company
- Guangzhou Shipbuilding Company
- Jiujiang Shipbuilding Company

The Shanghai Shipbuilding Company operates the four main shipbuilding centers in the region:

- Jiangnan Shipyard (Group)
- Shanghai Waigaoqiao Shipbuilding Company
- Hudong-Zhonghua Shipbuilding Group
- Shanghai-Chengxi Shipbuilding Company

The Guangzhou operations, run by the Guangzhou Shipbuilding Company, include:

- Guangzhou Shipyard International Company
- Guangzhou Huangpu Shipbuilding Company
- Guangzhou Longxue Shipbuilding Company
- Guangzhou Wenchong Shipbuilding Company

Shanghai Shipbuilding Company divisions

Jiangnan Shipyard (Group) was founded in 1865. The Jiangnan shipyard in Shanghai has always been, and still remains, the main supplier of domestically built warships to the Chinese Navy. In 1950, the shipyard was retooled with Soviet technical assistance. In 1996, the shipyard became the core asset of the Jiangnan Shipyard (Group) Co., which also took over the Quixin Shipyard in Shanghai in 2000, as well as several machine-building and ship repair plants.

In 2005, the company launched a project to relocate its main facilities from the old shipyard on the Yellow River to a new site outside the Shanghai city limits, on Changxing Island in the Yangtze delta. The new shipyard was named Shanghai Jiangnan Changxing Shipbuilding Company, and then renamed Shanghai Jiangnan Changxing Shipbuilding and Heavy Industry Corporation. In the summer of 2011, the new shipyard was incorporated into the "old" Jiangnan Shipyard (Group). In May 2011, Japan Shipbuilding Corporation acquired a 65% stake in Shanghai Jiangnan Changxing Shipbuilding and Heavy Industry Corporation (i.e., the new main shipyard). All construction work at the site is scheduled for completion by 2015; Changxing Island will host the largest shipbuilding complex in China and indeed the whole world, capable of building 4.5 million GT of commercial ships every year. The cost of the project is estimated at $3.6 billion. Shipbuilding operations at the new site commenced in 2009.

The old Jiangnan Shipyard site had two building berths, 275 m and 242 m long; four small slips, and three shipbuilding docks (232 m, 187 m and 147 m long). These facilities were used for both commercial and military shipbuilding. The new site on Changxing Island will have four dry shipbuilding docks (the two largest are 520 x 76 m and 510 x 106 m in size), plus 17 assembly floors and covered building berths.

In 1957, Jiangnan Shipyard built China's first submarine (a Project 613 boat, Whiskey class, assembled from Soviet components). Later on, it launched mass production of Type 033 (Project 633, Romeo class) subs. The shipyard resumed production of submarines in 2000 after a long break, and in 2004 launched a Type 039G1 boat. It has built two more submarines of the same type since then.

Jiangnan Shipyard took part in the program to build Type 051 (Luda class) fleet destroyers and Type 053H (Janghu and Jangwei class) frigates. In the 1990s, it built two Type 052 fleet destroyers (Luhu class). After 2000, it launched production of the new Type 052B (Luyang I class) and 052C (Luyang II class) destroyers. From 2004 to 2006, it delivered two Type 052B and two Type 052C ships to the Chinese Navy; two more Type 052C destroyers are now on the way. The company seems likely to retain its leading role in the production of China's largest and most advanced warships. Also, Jiangnan Shipyard had previously built several large and technically complex auxiliary ships, including three Dajing class submarine depot ships and six Yuanwang class space tracking ships (the last two of which were launched in 2005-2006). In 2009, the shipyard launched a Dongdiao class reconnaissance ship. There have also been reports that the new shipyard on Changxing Island will build domestically developed aircraft carriers.

Quixin Shipyard, which is part of the Shanghai shipbuilding group, was founded in 1902 as a commercial venture. Under the Communist government the shipyard became one of the leading Chinese manufacturers of attack boats and small warships, although now civilian contracts once again make up the greater part of its business. In 1998, Quixin Shipyard launched mass production of high-speed boats with aluminum hulls. In August 2000, the company was incorporated into the Jiangnan Shipyard (Group). Quixin Shipyard now has several covered berths, three open building berths (135 m, 70 m and 60 m) and a shipbuilding dock (90 x 18 m).

Previously Quixin Shipyard was the main producer of Type 037 (Hainan class) anti-submarine boats and Type 062 (Shanghai class) patrol boats. It appears that the company continues to produce small numbers of modified boats of the two types for export. It has also built seven Type 082 (Wosao class) harbor minesweepers. In 2005, it delivered to the Chinese Navy a Wozang class ship, which was an experimental new-generation mine countermeasures vessel. Later on, it launched production of a new Type 081 (Woichi class) minesweeper, of which at least five have already been built. Starting in 2004, Quixin Shipyard also became the main producer of Type 022 (Houbei class) wave-piercing catamaran missile fast attack craft with aluminum hulls. Several other shipyards also make Houbei class craft, but most of the 40-plus boats delivered to the Chinese Navy by 2011 were made in Shanghai.

Quixin Shipyard also builds assault landing hovercraft. In 2008, it launched a new Yuyi class large air-cushion landing craft, similar to America's LCAC craft.

The auxiliary vessels built by Quixin Shipyard include the *Shichang*, a 10,000 GT helicopter carrier delivered to the Navy in 1997 and used mainly as a training ship.

Shanghai Waigaoqiao Shipbuilding Company was formed in 1999 to build large new shipyard facilities near Shanghai. It took over some of the assets of Jiangnan Shipyard. The company is partly owned by the Japanese. It has built China's largest shipyard on Changxing Island in the Yangtze delta near Shanghai, and another shipyard for offshore drilling projects. The group is considered to be the leader of the Chinese shipbuilding industry. The facilities it operates will reportedly be able to build 7 million GT of ships every year by 2015. The group includes Shanghai Jiangnan Changxing Shipbuilding and Heavy Industry Corporation (the new shipyard on Changxing Island), Shanghai Lingang Offshore & Marine Company (offshore drilling projects), and Shanghai Xinye Marine & Engineering Design Company.

Hudong-Zhonghua Shipbuilding Group was formed in 2000 through a merger of two Shanghai shipyards—Hudong Shipyard and Zhonghua Shipyard. It employs more than 15,000 people.

Hudong Shipyard was founded in 1928. In the 1950s, the Soviet Union helped to turn it into a powerful shipbuilding operation. In 1955, it laid down two Soviet-designed (Project 50, Riga class) frigates. It remains one of the leading producers of frigates for the Chinese Navy, having built more Type 053H (Janghu and Jangwei class) ships of various modifications than any other shipyard in China. In the 1960s and 1970s, Hudong Shipyard built large numbers of Type 025 and Type 026 (Huchuan class) hydrofoil motor torpedo boats, as well as various auxiliary ships. Some of the frigates it has built were destined for export. In 2005, it delivered to the Chinese

Navy the first new Jiangkai I class (Type 054) frigate. Later on, it launched mass production of modified Type 054A (Jiangkai II class) frigates. Four have already been built, and two more have been laid down. In 2004, it built the first Fuchi class auxiliary replenishment ship, and in 2006, a Dahua class large survey and research ship. It has also delivered three Type F-22P frigates to Pakistan and two patrol ships to Thailand.

Hudong Shipyards is also a large manufacturer of commercial ships. Its Hudong Heavy Machinery subsidiary makes diesel engines, mostly under Western licenses; the engine division's combined annual output is 250,000 kilowatt-hours. The company also makes shipping containers. Since the 1980s, its shipbuilding facilities have undergone extensive retooling programs. They now include a 380 x 92 m dry shipbuilding dock, two large building berths and eight small ones. In 1996 Hudong Shipyard became the core of the newly formed Hudong Shipbuilding Group, which incorporated Zhonghua Shipyard in April 2000 to become Hudong-Zhonghua Shipbuilding Group.

The former Zhonghua Shipyard, which was founded in 1926, is now called Hudong-Zhonghua Shipyard. It has become a large commercial shipbuilding company. Its facilities include three building berths, which are 350 m, 185 m and 145 m in length. It employs up to 4,000 people. The shipyard has traditionally been the key maker of landing ships for the Chinese Navy. It has built 10 Type 072-II (Yuting I class) large tank landing ships since 1992. In the past decade, it has delivered three modified Type 072-II (Yuting II class) ships and three Yunshu class medium tank landing ships. In the past, the company built survey ships, and in 1997 it launched a 6,000 t Dahua class trial ship, which is used for testing modern naval weapons systems.

Hudong-Zhonghua Shipyard also builds Type 071 (Yuzhao class) amphibious landing platform docks. The first ship in the series, *Kunlunshan*, was delivered in 2007; it is the largest ship in service with the Chinese Navy. A second Type 071 ship was completed in 2011, and a third has been laid down.

In 2005, Hudong-Zhonghua Shipbuilding Group finished the construction of a new shipbuilding and ship repair base on Chongming Island on the Yangtze, near Shanghai. The base includes the new Shanghai Shipyard, which has a 300 x 46 m building berth, two dry shipbuilding docks (262 x 44 m and 205 x 36 m), and three floating docks. Huarun Dadong Dockyard, a large ship repair company based on the same island, has a huge floating dock that can accommodate ships of up to 200,000 GT. It also has a large dry dock and three smaller floating docks.

Shanghai-Chengxi Shipbuilding Company was formed in 2004 in Shanghai by consolidating the old Shanghai Shipyard facilities (in operation since 1862) and Chengxi Shipyard. The group builds and repairs commercial ships and produces some naval equipment. It has two dry docks, 262 m and 205 m long, a large building berth and three large floating docks. It is currently building a large new dry shipbuilding dock, which will enable the company to increase its annual production capacity to 1 million GT. In previous years, the old Shanghai Shipyard facilities were used to build tugboats for the Chinese Navy. In 2004, the shipyard built a Yubei class small tank landing craft.

Guangzhou Shipbuilding Company divisions

Guangzhou Shipyard International Company was founded in 1954 with Soviet assistance as the Guangzhou Shipyard. In 1955, it laid down two Soviet designed frigates (Project 50, Riga class). Until the early 1990s, Guangzhou Shipyard (also known as Donglang Shipyard) was one of the leading manufacturers of warships in China. It built Type 051 (Luda class) destroyers, Type 010 (T-43 class) minesweepers, fast attack missile boats and auxiliary ships. It also built large numbers of civilian ships, becoming the first Chinese shipbuilder to secure export contracts. In 1993, the shipyard was one of the first Chinese companies to become a joint venture with foreign capital, the Guangzhou Shipyard International Company. CSSC now owns only a 35.7% stake in the operation. Also in the 1990s, the shipyard underwent a major retooling program. Its facilities now include three building berths (277 m, 255 m and 190 m) and two dry docks, making it the largest shipbuilder in southern China. At present, military contracts make up a very small part of the company's business. Most of its revenues come from building civilian ships for export. Nevertheless, since 2000 it has been building large auxiliary ships for the Chinese Navy. In 2004, it launched the second Fuchi class replenishment ship; in 2007, a Danyao class transport and a Dalao class submarine depot ship; in 2008, an Anwei class hospital ship (another one is on the way); and in 2009, a submarine rescue ship. The company also builds large patrol ships for China's various paramilitary agencies. In 2005, it delivered the Haixun 31, a 3,300 GT large patrol ship, to the China Maritime Safety Agency.

Guangzhou Huangpu Shipbuilding Company was founded in 1845. In the 1950s and 1960s, the shipyard, known back then as Plant No. 201, underwent a radical retooling program with Soviet assistance. It used to build Type 033 (Romeo class) submarines, Type 053H (Janghu and Jangwei class) frigates of various modifications, and large numbers of attack boats. According to Western sources, in the 1980s China had plans to make the Huangpu Shipyard its second nuclear-powered submarine production center, after Bohai. But due to the extremely sluggish pace of the Chinese program to develop new-generation nuclear-powered submarines, those plans were eventually cancelled. Starting in 2000, the company (along with the Hudong Shipyard in Shanghai) has been building Type 054 and 054A (Jiangkai I and II class) frigates for the Chinese Navy. Four of the ships have already been launched; two more are on the way. The company also builds large patrol ships and boats for China's paramilitary agencies, as well as high-speed boats, including Type 022 (Houbei class) missile boats.

Like all the other large Chinese shipyards, Guangzhou Huangpu Shipbuilding has a large civilian business. It is also a large ship repair center, having been designated as the main ship repair base for the Southern Fleet. It has three large building berths (two of them 188 m in length, and one 130 m), one smaller berth, a 115 m dry dock and two large floating docks. The whole facility occupies an area of 1.1 sq.km. and employs 15,000 people.

Guangzhou Longxue Shipbuilding Company is a new commercial shipyard under construction since the mid-2000s on Longxue Island, 70 km from Guangzhou. The company's annual shipbuilding capacity has already reached 2.12 million GT, and is expected to rise to 3.5 million GT once all its facilities have been launched. The first ship was laid down in 2008. The shipyard has two 400 m long dry shipbuilding docks equipped with four 600 t cranes.

Guangzhou Wenchong Shipbuilding Company is a shipbuilding and ship repair company that was created in the 1980s and 1990s. It employs 4,500 people. It has three dry docks (300 m, 247 m and 202 m) and a 178 m building berth.

In addition to the Shanghai and Guangzhou Shipbuilding Company assets, the CSSC *Jiujiang Shipbuilding Company* group has a number of shipbuilding assets in Guangxi Province. These assets include several outfits involved in military shipbuilding, such as *Xijiang Shipbuilding Company* in Liuzhou, which employs 1,400 people and builds small- and medium-sized civilian ships, as well as small auxiliary ships for the Chinese Navy, coastal patrol boats and patrol ships.

Companies involved in military shipbuilding but not affiliated with CSIC or CSSC

Huanghai Shipbuilding Corporation, located in Weihai (Shandong Province), was founded in 1944. Until 2007, it was known as Shandong Huanghai Shipbuilding Company. In 2008-2009, it underwent an ambitious retooling and modernization program. It employs 2,200 people and has three large building berths plus three smaller ones. The bulk of its business is in the civilian sector, but in 2009 it delivered the *Haixun 11*, a 3,500 GT large patrol ship, to the China Maritime Safety Administration.

Wuhu Xinlian Shipbuilding Company, located on the Yangtze, was built in the late 1950s and radically modernized in the 1990s. Most of its business is in the civilian sector. It employs 3,500 people. The old facility in Wuhu has a 190 x 28 m building berth, six open assembly floors and a slip. In 2008, the company started the construction of a new facility, the Sanshan New Shipyard. It will have two 240 m building berths and one 220 m berth, with an annual production capacity of up to 1 million GT. The company also has a fiberglass shipbuilding division. In the past, Wuhu Xinlian Shipbuilding mass-produced Type 024 fast attack missile boats and Type 074 small tank landing ships. In 2004, it also built two Yunshu class medium tank landing ships.

Hangzhou Dongfeng Shipbuilding Company is located in Zhejiang Province and is controlled by the local government. It builds commercial ships and employs 820 people. It has two dry docks, eight building berths and one slip. It also builds Type 208B patrol boats for the Chinese police.

Zhejiang Shipbuilding Corporation is now part of the SINOPACIFIC Shipbuilding Group. After retooling from 2003 to 2010, it became an important player in the commercial shipbuilding sector. Previously it used to build fast attack missile craft and auxiliary ships for the Chinese Navy. In 2004, it launched a Yubei class small tank landing ship, and in the 1990s it built several Houdong class missile boats for Iran.

Tongfang Jiangxin Shipbuilding Company, located in Jiujiang (Jiangxi Province), now belongs to the Tsinghua Tongfang Corporation. It builds medium-sized commercial ships, as well as small auxiliary ships and tugboats for the Chinese Navy and police. Previously it used to build landing craft and patrol boats.

Taizhou Wuzhou Shipbuilding Industry Company, a former municipal shipyard in Taizhou (Zhejiang Province), has now been privatized. Previously it built fishing vessels, but after retooling in recent years, it now makes medium-sized commercial ships and small auxiliary ships for the Chinese Navy. It has two dry docks and two building berths.

Guangxi Gujiang Shipyard Company, located in Wuzhou (Guangxi Province), builds commercial ships. Some of the vessels it builds have aluminum hulls. It has made several patrol boats and small tankers for the Chinese Navy.

Two private shipyards in Guangdong Province, *Zhuhai Chenlong Shipyard Company* and *Shenzhen Hispeed Boats Technology and Development Company,* build small boats for the Chinese military and government agencies.

The Chinese Navy also operates six ship repair centers. Formally they are subordinate to the Central Military Council's Main Department for Weapons and Military Equipment. In addition to repairing the Chinese Navy's ships, these shipyards are sometimes used to build small ships, boats and auxiliary vessels.

The Navy ship repair center in Qingdao used to build patrol boats. In the past decade, it has built two Yunshu class medium tank landing ships, and, according to some reports, four Type 074A (Yubei class) ships. Now it builds Type 022 (Houbei class) missile boats and patrol boats.

The Navy ship repair center in Lushun (formerly Port Arthur) launched three Yunshu class medium tank landing ships in 2004. In recent years this shipyard, as well as the shipyard in Dinghai, launched at least two Yubei class (Type 074A) small tank landing ships each.

Conclusion

The Chinese shipbuilding sector has shown unprecedented growth in the past 15 years. Over a relatively short period, China has built huge and fairly advanced new shipyards, the biggest in the world. Its progress in building large ships in dry docks has been nothing short of astounding. Since 2000, some 22 dry docks of over 300 m in length have either already been built or are nearing completion, including six gigantic docks 480 m long or longer. At present, China has more dock capacity than any other country in the world, enabling it to build large numbers of commercial and military ships of any size or type. To illustrate, Russia does not have a single dry dock over 300 m in length.

Over a very short period, China has created a huge and fairly advanced commercial shipbuilding industry, which can be converted to military uses with relative ease. Theoretically, that industry can build any number of warships China might require in the foreseeable future. To all practical purposes, the Chinese Navy already has unlimited shipbuilding capacity at its disposal. The limiting factor now is not the capacity, and not even financial resources, but the fact that China is still lagging behind the world leaders in a number of crucial technologies, including naval weapons, electronics and propulsion. In these areas the Chinese Navy still depends on imports of key systems and components (including gas turbines from Ukraine, and torpedoes, SAM systems, radars and deck helicopters from Russia), or has to make do with obsolete domestic solutions (hydroacoustics, mine countermeasures). It is this technological gap that China needs to close as it starts building larger and more advanced new-generation warships in the years to come.

1.5 Case Study: Aircraft Engines

A leading Chinese weapons designer once said that the situation with engines had for a long time been "a disease crippling the entire Chinese defense industry, from aerospace to shipbuilding."[51] The inability to produce a modern aircraft engine still remains the key obstacle to the growth of Chinese military and civilian aerospace capability. No wonder then that the AVIC state aerospace corporation views engines as a top priority for investment. It continues to consolidate Chinese assets in this sector under a single specialized subsidiary it controls—AVIC Aero-Engines. AVIC is also making efforts to acquire aerospace engine assets abroad.

China is now working on a broad range of advanced new-generation engines, from piston motors for UAVs to a turbofan engine for the Y-20, a future heavy transport. A task that is critically important for national security and arms exports growth is to develop and/or improve domestic replacements for the main types of engines now sourced from Russia, including the AL 31F/FN, RD-93, and D-30KP2. This would enable China to increase its fleet of the advanced J-10, J-11B/BS and H-6K aircraft and to export them (along with the FC-1 fighter jet already being offered to foreign customers) without any Russian-imposed restrictions. In essence, that is the last major obstacle facing China on its path to becoming a leading arms exporter and a world-class military power.

General situation in the sector

The Chinese aircraft engines sector has made significant progress over the past decade. Previously it was barely able to assemble Soviet or Western engines under license, or to make very minor upgrades of foreign technology. But since then it has launched mass production of two domestically developed turbofan fighter engines: the Kunlun (used on the J-8II and J-7 families of aircraft) and the Taihang (used on the J-11B/BS and at some point in the future on the J-10). It has also finally managed, after trying since the early 1970s, to launch production of the Rolls-Royce Spey RB.168 Mk 202 turbofan engine. It is now being made in China under the local designation WS-9. There are also plans for its future modernization. Thanks to this success, the Chinese defense industry has been able to launch mass production of the badly needed JH-7A tactical bomber.

These and several other achievements have brought about a certain degree of euphoria among the captains of the Chinese aerospace industry. The Chinese media are brimming with upbeat reports about the engine industry's progress. If some of the local media outlets and engine designers are to be believed, China has already achieved the 1980s level of Western engine technology, surpassing the Soviet Union and present-day Russia. They say that now just one final push is needed to catch up with the very best American engines.

The fact is, however, that China is forced to continue buying AL-31F/FN turbofan engines from Russia. The latest known contract in June 2011 was for 123 AL-31FN engines, worth about $500 million.[52] Earlier in the year, there was another contract for 150 AL-31F engines for the Chinese fleet of Su-27, J-11 and Su-30 fighters. Indeed, Chinese orders have been growing so rapidly that the maker of these engines, the Salyut plant in Moscow, has no spare capacity left and is having trouble procuring the necessary materials. Yet another contract for 140 AL-31FN engines was expected to be signed in October 2011.[53]

Mass production of the Taihang engine is already under way, with several modifications in the pipeline. But refitting the entire fleet with this engine does not make sense for China, given that it is now peacetime and supplies of other engines from Russia are available. A wholesale transition to domestically made engines would inevitably lead to more air accidents and higher operating costs. The new Chinese engines have a relatively short lifespan, in addition to numerous design flaws and manufacturing problems that have yet to be eliminated. This, however, is the best China has been able to achieve so far with its largest, most advanced and most important aircraft engine program.

Meanwhile, the projects to develop domestic versions of the RD-93 and the D-30KP2 turbofan engines, known respectively as the WS-13 Taishan and the WS-18, are even less mature. This is true despite some reports claiming that the Taishan engine has been successfully tested on the FC-1 fighter and has entered mass production.

Export versions of Chinese aircraft (such as the Z-9 helicopter and its various modifications, the K-8 trainer, and new versions of the Y-7/8/12 transports) are almost universally fitted with imported engines, even though domestic options are available. The entire civilian aircraft export program relies on imported engines; this applies to helicopters as well as planes.

Whenever there is such a possibility, AVIC tries to close the technological gap by acquiring technologies and entire companies abroad. The largest such deal has been the acquisition in December 2010 of Continental Motors, a large US manufacturer of aircraft piston engines.[54] The Chinese paid $186 million for the company. AVIC said it did not intend to make any serious changes to the company's operations or to wind down production in the US. It did say, however, that it would be sending Chinese specialists to Continental Motors facilities in the US. Meanwhile, AVIC Commercial Aircraft Engine Company, set up with the specific purpose of developing civilian engines, is hiring large numbers of foreign specialists and signing cooperation agreements with leading foreign universities and research centers.

Industry structure

The key Chinese aircraft engine assets are owned by AVIC; they are now undergoing an extensive program of reform and restructuring. The bulk of the industry is being consolidated under AVIC Aero-Engines. Another division specializing only in civilian engines is AVIC Commercial Aircraft Engine Company (ACAE). Its reason for being is to achieve a breakthrough in civilian engine technology by making the best possible use of foreign partners and specialists hired from abroad.

AVIC Aero-Engines

AVIC Aero-Engines was set up on September 16, 2010. It has consolidated assets in such areas as aircraft engines, generators, gas turbine engines, and helicopter gearboxes. According to an AVIC press report released in March 2011, the holding will eventually include 23 production companies and research facilities.[55]

The company's authorized capital is a mere 10 million RMB ($1.5 million); it is registered in the Beijing suburb of Shunyi.[56] Organizationally it is a unitary company,[57] meaning that eventual flotation is not in the cards. It does, however, control several listed companies, including Xian Aero-Engine PLC and Chengdu Aero-Engine. AVIC Aero-Engines' sole shareholder is AVIC itself.

The company does not have any production operations of its own; it merely controls the assets consolidated under its umbrella. The reason for setting it up in the first place was to improve coordination and management standards in the sector, as well as to oversee investment operations and R&D financing.[58]

The internal structure of the holding is still being shaped. On June 22, 2011, AVIC decided to make an unpaid transfer of shares in four companies, including Xian Aero-Engine PLC,[59] to AVIC Aero-Engines.[60] AVIC Aero-Engines was given[61] an 83.47% stake in Xian Aero-Engine Group Co., and a 52.85% stake in Chengdu Aero-Engine, plus full ownership of Beijing Changkong Machinery Company.[62]

The move essentially meant that AVIC Aero-Engines was also given control of the AVIC Aero-Engine Controls (AAEC) holding. AAEC owns the suppliers of important aerospace components, including automated command and control systems. AVIC Aero-Engines is now the largest AAEC shareholder, with a 45.31% stake. Another 35.22% of the shares are held by other AVIC divisions, and the remaining 19.47% are owned by small shareholders and traded on the stock market.[63]

Nevertheless, as of June 2011, consolidation of the sector under AVIC Aero-Engines was far from complete. For example, it has yet to become the formal owner of shares in such key industry players as China National South Aviation Industry Co., which remains a direct subsidiary of AVIC.[64] Neither have there been any reports about Shengyang Liming, a giant of the Chinese engine industry, being made part of AVIC Aero-Engines. Legal procedures for the transfer of some other engine industry assets are still under way. The process will be long, owing to the complicated structure of the industry. But for all intents and purposes, the decision to consolidate Chinese aerospace engine assets under AVIC Aero-Engines has already been implemented, and the company itself is up and running.

AVIC Commercial Aircraft Engine Company

Another major industry player is AVIC Commercial Aircraft Engine Company (ACAE),[65] which specializes in engines for civil aviation. ACAE was set up on January 18, 2009; its authorized capital is 6 billion RMB ($940 million). It is a three-party joint venture between AVIC; Shanghai Electric Group, which is China's largest maker of power generation and industrial equipment; and Shanghai Guosheng, an investment corporation owned by the Shanghai city government. AVIC holds 40% of ACAE shares; Shanghai Electric, 15%; and Shanghai Guosheng, another 15%.[66] The owner of the remaining 30% has not been disclosed. It appears that like many other AVIC subsidiaries, ACAE will eventually be floated, so the 30% may have been reserved for an IPO.

The key objective set for ACAE is to produce a turbofan engine (and its subsequent modifications) for the C919, China's first long-haul jet liner being developed by COMAC, a commercial company not owned by AVIC. The current plan is that the C919 will enter a flight testing program in 2014 and commercial service in 2016. The first units will be equipped with the LEAP-X1C turbofan engine, made by a consortium of GE and Snecma CFM International. The engine's thrust is 111.2-133.4 kilonewtons.

Later, however, it will be replaced by the domestic CJ-1000A engine, which ACAE is now developing. The integrated project group in charge of the program was formed as recently as June 28, 2011.[67] In September the CJ-1000A engine was given the Yangtze designation.[68] On earlier occasions, it was called the SF-A. The developers have already demonstrated its mock-up. The engine's precise specifications have not been disclosed, but clearly they must be in the same range as the LEAP-X1C engine. The chief executive of ACAE, Wang Zhi Lin, said in April 2011 that shipments could commence after 2020, and that the engine's thrust will be between 120-130 kN.[69]

ACAE is now working to build new R&D and manufacturing facilities in Shanghai, establish cooperation with foreign partners and identify subcontractors. It appears that the company started to recruit in large numbers only in August 2010, when the first 300 staff were hired at a special ceremony (which involved taking an oath to do everything humanly possible for the development of a new Chinese engine).[70] ACAE has also begun to purchase software: It has already rolled out an enterprise PDM system supplied by EPICCA (a report to that effect was released in June 2011).[71]

Unlike AVIC Aero-Engines, which largely relies on existing Chinese R&D and manufacturing capability, ACAE aims to develop everything from scratch. Its strategy includes attracting private capital, hiring large numbers of foreign specialists and establishing an extensive network of international partnerships. On the whole, the ACAE model is similar to the one used by COMAC—but whereas COMAC is no longer controlled by AVIC, ACAE remains part of the Chinese aerospace monopoly.

In May 2010, ACAE rolled out a long-term (three- to five-year) global recruiting program. It aims to hire up to 2,000 specialists in 79 different areas, with at least five years of experience in the industry.[72] For now, however, there is no information about the program's progress or about the performance of those specialists who have already been hired.

It must be said, however, that state-owned Chinese companies in the civilian sector have had little positive experience in bringing in foreign experts and managers. The foreigners almost always find it difficult to adjust to the local business and cultural specifics, or to establish productive collaboration with their Chinese colleagues. Early termination of contracts is nothing out of the ordinary. Putting the foreign specialists to good use will therefore be a serious challenge for ACAE, regardless of how attractive the compensation packages are.

Another important priority for ACAE is to establish contacts with leading research centers both in China and abroad to outsource some of the work and pursue more effective staff support policies. The company is known to have already signed cooperation agreements with several Chinese universities and at least one in Britain. It is also likely to pursue a program of establishing research and technology centers abroad. It has already signed initial memorandums of intention with several Western companies. One of the first such companies was Germany's MTU.[73]

Aircraft engine R&D and manufacturing assets

Harbin Dongan Engine Manufacturing Company (DEMC, a.k.a. Plant No. 120)[74]

Founded in 1948 as a repair shop, DEMC is probably China's oldest company in the sector. For a long time, its main product was the HS-7 piston engine (a copy of the Soviet ASh-82V) for the Z-5 helicopter (a copy of the Soviet Mi-4) and its later modifications. Later on, the company launched production of the WJ-5 turboprop engine (a version of the Soviet AI-24) used on China's main air transport workhorse, the Y-7 (An-24), made in Xi'an. DEMC makes propeller gearboxes for all three of the most widely used helicopters in China—the Z-8, Z-9 and Z-11.[75] It also supplies various other components, including power generators and fuel pumps.[76] The company's other important line of business is gas turbine power plants, including mobile units.[77] Their production is set to increase dramatically following the Chinese government's decision to build more gas-fired power plants using gas turbine technology. DEMC has its own R&D center which employs more than 1,000 staff, including 700 engineers.[78]

Shenyang Liming Aero-Engine Group Corporation (LMAEG, a.k.a. Plant No. 410)[79]

LMAEG evolved from one of China's oldest defense companies—the Northeastern Arsenal, founded in 1919 in Shenyang to supply weapons for ground troops. In March 1954, when China launched production of its first fighter aircraft in Shenyang, the company completed a retooling program and started making aircraft engines. In 2001, LMAEG, which was then part of the Shenyang Aircraft Industry Corporation (a subsidiary of what was then the AVIC I aerospace holding company), became a top-tier AVIC I subsidiary.

The company's facilities occupy an area of 2.99 million square meters. It employs 16,000 people, including 1,000 engineers, and has about 16,000 machine tools. Its revenues reached 6.6 billion RMB ($1 billion) in 2009; its employees earn an average of 45,000 RMB ($6,590) per annum.[80] That is more than double the average for the city of Shenyang (18,560 RMB in 2009).[81]

Engines made by LMAEG are designed by the Shenyang Engine Design and Research Institute (SAERI, a.k.a. Institute No. 606). This is the largest research facility of its kind in China. Both of China's domestically developed jet engines, the Kunlun and the Taihang, were designed by SAERI. Since its foundation in 1961, the institute has produced 11 jet engine designs. It employs 2,200 people, including 1,000 engineers, of whom some 400 hold doctoral degrees.[82] At least until recently, the institute was directly subordinate to AVIC, as opposed to being part of the LMAEG corporation.

LMAEG was the birthplace of the Chinese jet engine industry. It not only manufactured China's first turbojet engine (the WP-5, a copy of the Soviet VK-1, used on the J-5 fighters) but also the country's first liquid-propellant rocket engines. LMAEG also made the WP-6 and WP-7 engines (versions of the Russian R-9BF-811 and R-11-30 engines for the J-6 and J-7 aircraft, respectively) and their various modifications. It was involved in rolling out mass production of China's first high-power turbojet, the WP-8 (a copy of the Russian RD-3M, made in Xi'an and used on the H-6 bombers). At present its product range includes the WP-15 Kunlun turbojet engine (used on the latest modifications of the J-7 and J-8II fighter jets) and the WS-10A Taihang turbofan engine (used on the J-11B, J-11BS and in the future, the J-10). The company now appears to be focusing on improving the performance of the WS-10A family of engines so that they can be rolled out for the entire J-11 and J-10 fleets. Another priority is the WS-10G turbofan engine, presumably being developed for the fifth-generation J-20 fighter.

China National South Aviation Industry Company (CNSAIC, a.k.a. Plant No. 311)[83]

Founded in 1951, CNSAIC is now the main Chinese manufacturer of turboshaft aircraft engines, used mostly on helicopters. It used to be an independent state-owned company. Until 2006 or even later, it was directly subordinate to the government's SASAC.[84] CNSAIC has a diverse product range; much of it is civilian and non-aerospace. It plays a role in the production of piston and turbofan engines. It has about 5,300 units of modern manufacturing equipment[85] and occupies an area of 5.5 sq.km.

CNSAIC has 16 subsidiaries, including a division that makes motorcycles and motorcycle engines. The company manufacturers the HS-6 piston engine used on the Y-5 (An-2) light transports. Its annual production capacity for the HS-6 is 200 units.[86] It also makes the WJ-6 engine used on the Y-8 medium transports (a version of the Soviet An-12), with annual production capacity of up to 100 units.[87]

For the helicopter segment of the market, CNSAIC makes the WZ-8A turboshaft engine, a licensed copy of the French Turbomeca Arriel 1C1, used on the Z-9 helicopter (a version of the Eurocopter AS365 Dauphin made under license). Due to military sensitivities, the production capacity for this engine is not disclosed.[88] The company also makes the WZ-8D turboshaft engine used on the Z-11 light helicopter; it is now working on more advanced modifications of that engine.[89]

CNSAIC is developing the WZ-9 turboshaft engine for the Z-10 attack helicopter. The Chinese have reportedly now decided to equip these helicopters with domestic engines (initially they sourced them from Pratt & Whitney),[90] but the rollout of the WZ-9 is behind schedule. This is likely due to the teething problems that every new engine has, and which usually take the Chinese years to get rid of. CNSAIC is also working on engines for the future localized Z-15 helicopter (EC175), a joint project between Avicopter and Eurocopter. AVIC is now working on a program to transform its CNSAIC division into China's main producer of engines for light and medium aircraft.

AVIC Xian Aeroengine (Group) Ltd. (XAEC, a.k.a. Plant No. 430)[91]

Founded in 1958, the plant now has 4,500 units of manufacturing equipment and 2,500 engineering and technical staff. Unlike the AVIC companies discussed above, in 2001 XAEC transferred all its assets to a publically traded subsidiary, Xian Aeroengine PLC. As a result, there is quite a lot of information available about the group's economic indicators. Its corporate report for the first half of 2011 is available on the Shanghai Stock Exchange website.[92]

In the reported period, the company's net profit rose to 138.565 million RMB ($21.8 million), up 34.22% from the same period last year. Revenues from sales of aircraft engines were up 25.52% to 1.748 billion RMB ($275 million). Revenues generated by non-aerospace sales were 777.661 million RMB. Exports, at 592.247 million RMB, accounted for 33.8% of the group's revenues. In 2010, the group made a net profit of 254.3 million RMB on revenues of 6.085 billion RMB.[93]

The company's key product is the WS-9 turbofan engine and its modifications. The WS-9 Qinling[94] is a licensed version of the Rolls-Royce Spey RB.168 Mk 202, which has been completely localized in China. The engine is used on the JH-7A, a tactical bomber made by the XAC corporation in Xi'an.

Previously XAEC used to make the WP-8 high-power turbojet for the H-6 bomber.[95] It is not clear whether that model is still in production. All the recently made H-6 family bombers are a radically upgraded H-6K modification equipped with Russian D-30KP2 turbofan engines. XAEC is also involved in the development of the WS-18 high-power turbofan engine, which should replace the D-30KP on the H-6K bomber and later be used on the heavy military transport the Chinese are now developing.

In 2010, XAEC took over another Chinese aerospace engine maker, the Guizhou Liyang Aero-Engine Company (a.k.a. Plant No. 460), which used to be part of the Guizhou Aircraft Industry Corporation, a maker of combat trainers and attack aircraft. The division used to make engines for the J-7 aircraft and for missile weapons. It also had a broad civilian product range. In 2009, it made 55.8 million RMB ($8.1 million) of net profit on 2 billion RMB ($292 million) of revenue.[96] Following the takeover, XAEC now controls another important Chinese aerospace engine program previously led by Guizhou Liyang Aero-Engine—the WS-13 Taishan turbofan engine (a copy of the RD-93). XAEC also supplies numerous components to other Chinese engine makers, including starters, turbine rotors, etc.

AVIC Chengdu Engine Group (CEG, a.k.a. Plant No. 420)[97]

Founded in 1958, the company has 4,700 employees and its assets are worth 4.07 billion RMB ($640 million). For a long time, its main products were engines used on various modifications of the J-7 fighters made in Chengdu. These included the WP-7 jet engine (a version of the Soviet R-9BF-811) and the WP-13 engine (a version of the Soviet R-13-300, made in cooperation with Guizhou Liyang Aero-Engine Company). The company is now involved in developing the WS-18 turbofan engine; indeed, according to some reports, it is the lead company in the program.[98] CEG is also a large supplier of components, including titanium parts, for other companies in the engine industry.

Key industry programs

WS-9 Qinling

Beijing bought the first batch of Rolls-Royce Spey RB.168 Mk 202 turbofan engines from the UK in 1972. In July 1973, it began negotiating the acquisition of a license to produce these engines in China. The contract was signed on December 13, 1975. XAEC was chosen as the local manufacturer of the engine, which received the Chinese designation WS-9. The first four units were assembled in China in the first half of 1979. In November, the engines underwent a joint program of tests, which involved running them for 150 hours. From February to May 1980, another series of tests was conducted in the UK, including a high-altitude and a low-temperature (below -40°C) test. The results were declared a success. But further efforts to launch mass production were hampered by the need to modernize several other sectors of Chinese industry and technology, including metallurgy, materials science and chemistry. It is said that as part of the WS-9 program, China had to develop 12 individual technologies that turned out to be world-class, and 46 others that were completely new for the country's industry. Until those related projects were completed, production of the WS-9 required imports of many materials and components. A series of standard bench tests (e.g., running the engine for 150 hours) of a 70% localized WS-9 began in November 1995. A project to achieve complete localization was launched in the second half of 1999. The 150-hour test was completed only in 2001. Flight tests began in June 2002. The development of a completely localized version was completed in July 2003. The engine then entered mass production and is now being used on the JH-7A bomber. The task of launching mass production of a localized version of the Spey engine was therefore accomplished some 31 years after the first British-made units were purchased, and 28 years after the Chinese acquired the license. The main reason it took so long was apparently the general technological backwardness of the Chinese industry in those days.

WP-14 Kunlun

The Kunlun is thought to be the first domestically designed Chinese aircraft engine to enter mass production. The completion of the project was announced in early 2002. Work on the WP-14 turbojet engine began in 1984 at Institute No. 606 in Shenyang after a series of failures that led to the cancellation of the WS-6 and WP-13 projects. But some of the solutions developed for those cancelled projects were used in the Kunlun. The first unit was assembled in 1987; bench tests began shortly afterwards. The engine entered a flight testing program on the J-8IIC fighter in December 1993. In late 1997, there was a crash caused by engine failure in mid-flight. The incident caused a long delay in the testing program. But by August 2001, some 658 test flights had been completed with the Kunlun. It was concluded that the engine could enter mass production once the flaws identified during the test program had been eliminated. The entire program to develop a domestic engine for the upgraded second-generation J-7 and J-8 fighters took about 18 years to complete. That is a long time by any standard, especially considering that the Chinese designers and engineers were able to use solutions already developed for previous projects, and to study very carefully Soviet engines such as the R-29 and R-13F-300 that they had obtained from Arab countries.

WS-10 Taihang family

The WS-10 turbofan engine family, designed for the fourth-generation Chinese fighter jets, has been in development since the early 1980s. The modifications produced at various stages of the program include the WS-10, WS-10A, WS-10B, WS-10C, WS-10D, and the future WS-10G, presumably meant for the fifth-generation J-20 fighter.

Work on the engine for the future Chinese fourth-generation fighters (Project 10) began at Institute No. 606 in 1984. The Chinese designers and engineers had obtained several units of Soviet and Western aviation engines, and used them as a starting point. Those engines included the CFM International CFM-56-II turbofan and the aforementioned Rolls-Royce Spey Mk 202. In addition, Egypt supplied one MiG-23 fighter equipped with the Soviet R-29-300 turbojet engine in exchange for a batch of J-6 fighters. The Chinese attempted to produce a clone of that engine, which they designated as the WP-15, but failed. The WS-10 made use of some solutions developed for another failed project, the WS-6G. Later on, in the 1990s, work on the

WS-10A version was influenced by the Chinese designers' intimate knowledge of the Russian AL-31F turbofan engines.

In 1986, it was decided to use the WS-10 for the future J-10 fighter. The Chinese hoped to achieve performance on a par with the American General Electric F110-GE-129 engine. The first WS-10 prototypes entered a program of tests in 1987 and completed it in 1993. By 1993, the Chinese were already considering the WS-10 both for the J-10 and for the Su-27SK aircraft imported from Russia. One Su-27 unit was acquired from Russia for the test program; the Chinese hoped that in the future they would create a localized copy of this aircraft. The project to adapt the WS-10 for use with the Su-27 began in December 1997. The first flight of the Su-27 with one of its two original AL-31F engines replaced with the Taihang took place in June 2002. Testing of the prototypes on the J-10 fighter began in 2002-2003 and was completed by late 2005. Sometime between 2002 and 2003, there was a serious accident during the tests. In 2006, the Chinese announced the completion of the WS-10A project. Judging from several available photos, in recent years the engine has been used on the J-11B/BS fighters serving with combat maneuver units of the Chinese Air Force. However, in 2009 senior AVIC executives admitted that there were problems with the engine's reliability, safety and performance. Finally, the best proof that the WS-10A is still struggling with teething problems is that China continues to buy large numbers of AL-31F/FN engines from Russia. The Chinese program to develop a reliable domestic engine for the fourth-generation fighters has borrowed extensively from Western and Russian designs, and made use of solutions developed for earlier Chinese projects. But 27 years on, it is still far from completion.

Conclusion

China still lacks an established domestic capability in aircraft engine making. The best that can be said about Chinese technology in this area is that instead of blind copying and minor upgrades of foreign designs, the Chinese have now begun to develop their own designs based on borrowed solutions. The captains of the Chinese engine industry know very well that there will be no tangible progress without broad international cooperation. The latest Chinese projects, which aim to develop advanced engines for civil aviation, rely heavily on working with world industry leaders, using foreign specialists, and buying technologies (or sometimes entire companies) abroad.

In the defense sector, however, the scope for international cooperation is limited, and here the Chinese aircraft engine makers are finding themselves in an especially difficult situation. Flight tests of the fifth-generation J-20 fighter, which began in 2011, indicate that the Chinese leadership is setting increasingly ambitious goals for the industry. These goals cannot be achieved by merely playing catch-up with the leaders and copying Western or Russian technology of 20 years ago. This explains why the Chinese are once again showing interest in cooperation with Russia in the area of aircraft engine making—and why Beijing is especially interested in the latest and most advanced Russian engine, known as Izdeliye 117S.

Chapter 2
China on the Arms Market

2.1 China's Position

Ever since the Communist government came to power in China in 1949, the country has constantly cycled between being a net exporter and net importer of weapons. During some periods, such as the 1950s, 1960s and 1992-2010, China has purchased more weapons than it sold. In doing so, the country managed to improve very significantly the overall level of its own defense technology and manufacturing capability. In the 1980s, however, China was one of the world's largest defense exporters. Its weapons were relatively low-tech—but they were also cheap, simple and easy to use. That helped Beijing win strong, and even dominant, positions in markets such as Egypt, Pakistan, Bangladesh, Thailand, Sri Lanka, and sub-Saharan Africa. During the Iran-Iraq war in 1980-1988, the two countries also bought lots of Chinese weapons.

After a period of being a large net importer in 1992-2005 (the second such period, after 1950-1962), the country has built up a formidable engineering and manufacturing capability to produce third- and fourth-generation weaponry. It is now well on its way to becoming a net arms exporter once again. In fact, it may have done so already.

To summarize, since 1949 China has gone through several import-export cycles, each lasting about 12 years. Each successive "net importer" period enabled Beijing to lay the industrial and technological foundations for the subsequent "net exporter" period.

The distinguishing feature of the most recent "net importer" cycle, which began in 1992, is the serious restrictions imposed by the West on weapons exports to China following the 1989 Tiananmen Square protests. The European arms embargo was imposed in June 1989. The US has also been restricting exports to China of advanced weaponry and technologies ever since the end of the cold war. As China continues its transformation into the world's second de facto superpower after America, these policies are set to become even more restrictive. Washington not only refuses to sell modern weapons to China, but also puts pressure on its allies and partners to follow suit. Israel recently yielded to US lobbying and refused to sell its Phalcon airborne early warning radars to Beijing. Israel was also forced to end cooperation with China on the J-10 project, a light fighter jet which is based on the Israeli IAI Lavi design.

As a result of these restrictions, Russia remained practically the sole supplier of modern defense technology to Beijing. Such dependence on just one big supplier has made it difficult for China to acquire the most advanced technologies. That became especially obvious at the turn of the century, when a second major batch of various weaponry that China had bought from Russia turned out to be surprisingly dated from a technological point of view. Using its near-monopoly, Moscow has limited the PLA's access to the most high-tech Russian weapons, or made that access conditional on Beijing buying in very large batches. In recent years, the Russian defense industry has been able to accumulate a large portfolio of contracts signed with Russia's own Ministry of Defense and with many foreign buyers. It is therefore safe to assume that China will now find it even more difficult to acquire advanced weapons technologies abroad. At one point, Beijing pinned great hopes on the prospect of the European Union lifting its arms embargo, which seemed quite realistic in 2004-2005. But the embargo remains in place to this day.

Another distinguishing feature of Chinese weapons imports is that China is keeping its spending on foreign weapons firmly in check. The country's financial muscle is already colossal and keeps growing, and the People's Liberation Army requires huge amounts of new weaponry. But Beijing is unwilling to spend too much on buying that weaponry abroad. In terms of absolute numbers of weapons and systems bought, including fighter jets, destroyers and submarines, China has been the biggest Russian defense customer since the end of the cold war. But the dollar value of those contracts is actually quite modest. In the early 1990s, much of the Chinese arms imports from

Russia were paid for in kind, with Chinese supplies of consumer goods. We estimate that over the past decade, Beijing has been spending an average of $2 billion a year on arms imports, though in 2005-2006 the figure may have peaked at $3 billion to $3.5 billion.

2.2 Russia on the Chinese Defense Market

Between the early 1990s and mid-2000s, China was one of the world's biggest arms importers and Russia's biggest defense customer. For a long time, Chinese contracts made up 40% to 45% of Russian arms exports, peaking at 60% in 2000. But by the mid-2000s, Chinese arms imports from Russia began to fall off sharply. Having made great progress in defense technology and manufacturing capability in the preceding 15 years, China started to rely mostly on its own defense industry to meet the PLA's requirements. In 2007, China was replaced by India as the largest importer of Russian weapons. At present the share of Chinese contracts in state-run defense exporter Rosoboronexport's order portfolio is 4% at the very most; Beijing is probably not even among the top five Russian defense industry customers.

Russian supplies in 1992-2010

After the breakup of the Soviet Union, China was the largest importer of Russian weaponry. Russia became the main source of advanced defense technology for China's rapidly developing and modernizing armed forces. We estimate that China was the destination for up to 40% of all weapons sold by Russia from 1992 to 2007. In 2004, some 57% of Russian weapons deliveries went to China. In 2005, when Beijing took delivery of six Project 636M (Kilo Mod class) submarines and a Project 956EM (Sovremenny Mod class) destroyer, that proportion may have been even higher. Only in 2007, well after the value of newly signed Chinese contracts had peaked, did China's share of Russian arms deliveries fall to an estimated 25% to 30%.

Known contracts and deliveries to China are listed in a table later in this section, but first, here is a brief summary of the main Russian defense exports to China.

In the aerospace segment:

- 38 Su-27SK fighter jets (in 1992 and 1996)
- 40 Su-27UBK combat trainers (in 1992, 1996 and 2000-2002)
- 76 Su-30MKK multirole fighters
- 24 Su-30MK2 fighters, equipped to take on naval targets, delivered to the Chinese Navy in 2004
- 95 Su-27SK kits for the assembly under license of Russian aircraft in Shenyang (1998-2003)

China's support fleet has 15 Il-76 transports supplied by Russia or built in the 1990s at the Tashkent Aircraft Plant in Uzbekistan. It also has four Il-78 aerial refueling tankers; their origin is unclear, but it is probably Ukraine. In 2005, Beijing placed an order for 34 new Il-76MD aircraft. Under the same contract, the Chinese Air Force was due to receive four more Il-78 flying tankers in 2005, but the deal fell through. China also placed an order for four A-50E airborne early warning aircraft. After Israel yielded to American pressure and refused to supply the Phalcon early warning radars for these planes, China was forced to develop its own system, the KJ-2000.

China has also been buying large numbers of dual-flow turbojet engines for the J-10 light fighter (Al-31FN turbofan engine) and for the FC-1 (RD-93 turbofan engine, more than 100 units sold). There were also contracts for the Al-31F engine, which was unrelated to the J-10 program; the engines were bought for the Su-27 and Su-30 aircraft whose original engines had reached the end of their lifespan.

Throughout the period from 1992 to 2010, Russia had been selling large numbers of helicopters to China. Its attempts to win Chinese contracts for the Mi-24/Mi-35 attack helicopters were unsuccessful, but the Mi-8 and Mi-17 transports have been a roaring success with the Chinese. The helicopter plants in Kazan and Ulan-Ude have made more than 300 of these machines under Chinese contracts. In 2007, China's Sichuan Lantian Helicopter began assembling Mi-171 helicopters in Wuhou from assembly kits supplied by the Ulan-Ude Aviation Plant.

Seven Ka-28 anti-submarine helicopters and three Ka-27PS search and rescue helicopters were supplied in 1993-2000 for the Chinese Navy.

In addition to helicopters, China has also purchased large batches of guided airborne weapons. The list includes air-to-air missiles such as the R-73E (AA-11, short range), R-27 (AA-10, medium range) and RVV-AE (AA-12, medium range); air-to-surface missiles such as the Kh-25 (AS-10), Kh-29TE (AS-14B) and Kh-59ME (AS-18); Kh-31P (AS-17) anti-radar missiles; Kh-31A (AS-17) and Kh-59MK (AS-18) anti-ship missiles; KAB-1500Kr guided bombs; and APR-3ME anti-submarine missiles (used with Ka-28 helicopters). Russia and China have launched joint production of the Kh-31P missile in China under the KR-1 (YJ-91) designation. Most of the missile's components, including the homing heads, are supplied by Russia's Tactical Missiles Corporation (KTRV). Russian companies are also providing assistance to China in the development of several guided airborne weapons, including the PL-12 (SD-10) air-to-air missile, for which Russia's MNII Agat has developed an active radar homing head.

In the *naval weapons and platforms* segment:

- Two Project 956E (Sovremenny class) fleet destroyers
- Two Project 956EM (Sovremenny Mod class) fleet destroyers
- Four diesel-electric submarines: two Project 877EK and two Project 636 (Kilo class) boats
- Eight Project 636M (Kilo Mod class) diesel-electric submarines equipped with the Club-S missile system
- Two Podsolnukh-E coastal surface wave effect radars

China has also purchased the Rif-M (SA-N-6/20) and Shtil-1 (SA-N-12) shipborne SAM systems for use with the Chinese Project 051C (Luzhou class) and 052B (Luyang I class) fleet destroyers, designed with the participation of Russia's Severnoye design bureau. Severnoye also appears to have been involved in the development of China's Project 052C (Luyang II class) fleet destroyers and Project 054 (Jiangkai I/II class) frigates. Meanwhile, the Rubin Design Bureau has been involved in designing several new Chinese submarines.

Apart from Russian-built ships, China has also received large numbers of naval weapons, systems and electronics. This includes anti-ship missiles such as the 3M80E Moskit (SS-N-22) and 3M54E Club (SS-N-27); the Shtil (SA-N-7) and Shtil-1 (SA-N-12) SAM systems; the Kashtan

(CADS-N-1/SA-N-11) gun/missile weapon system; the AK-130 twin-gun artillery system; the AK-630M 30mm six-gun artillery system; the TEST-71, 53-65K and SET-65E torpedoes, and other naval weaponry. In addition, Russia has supplied a number of weapons systems and radars for Chinese-built destroyers and frigates, including the aforementioned Rif-M and Shtil-1 (some of which were vertical-launch modifications), the Fregat-M2E and Mineral-ME radars, and other systems. Russia has already transferred to China a license for the AK-176 76mm artillery system. Finally, Russian companies have been developing torpedoes and mines for the Chinese Navy.

In the *air defense* segment:

- 12 battalions of the S-300PMU-1 (SA-20A) long-range SAM system
- 16 battalions of the S-300PMU-2 (SA-20B) long-range SAM system
- In 1996 and 1999, China took delivery of 27 Tor-M1 (SA15) short-range self-propelled SAM systems optimized for defense against high-precision weapons.

There have been indications that China has also received automated air defense command-and-control systems and radars. Finally, Russian companies that are now part of the Almaz-Antey air defense concern have been involved in the development of Chinese SAM systems, including the HQ-9 and several short-range systems.

The smallest segment by sales volume has been *ground weapons*:

- An estimated 1,000 2K25M Krasnopol-M 152/155mm guided artillery projectiles
- License for the 2K25M Krasnopol-M projectiles
- License for the Bakhcha-U gun turret and for the 9K116-3 Basnya 100mm gun-launched missile round armed with the 9M117M Kan (AT-10) guided missile, which is used with the Bakhcha-U
- License for the 9K119M Refleks-M 125mm rounds (armed with the 9M119M Invar, AT-11 missile) and the 9K116-1 Bastion 115mm rounds (armed with the 9M117M2 Arkan, AT-10 missile)
- License for the Shmel man-portable infantry flame-thrower
- License for the 2S23 Nona-SVK 120mm self-propelled mortar-gun (up to 100 of these systems may have been supplied to China)

- China is also thought to have acquired a license for the 9K58 Smerch 300mm MLR system (mounted on the 9A52 chassis) and for the gun turret of the 2S19M1 Msta-S self-propelled howitzer.

It should be stressed that, in the ground weapons segment, China has been buying not only finished products but also production licenses.

PRODUCT	GENERATION	SUPPLIER	CONTRACT SIGNED	DELIVER- IES	UNITS	VALUE	NOTES
Aviation systems and weaponry							
Su-27SK fighter	Fourth- generation fighter	Komsomolsk- on-Amur Aerospace Company (KnAAPO)	Presumably 1990	1992	20	Unknown, estimated at about $600 million	Up to one-third of the value paid in kind (consumer goods)
Su-27UBK combat trainer	Fourth- generation fighter	Irkutsk Aerospace Company (IAPO)	Presumably 1990	1992	6	Unknown, estimated at about $150 million	Up to one-third of the value paid in kind (consumer goods)
Su-27SK fighter	Fourth- generation fighter	KnAAPO	1995	1996	16	Unknown, estimated at up to $500 million	
Su-27UBK combat trainer	Fourth- generation fighter	IAPO	1995	1996	6	Unknown, estimated at $150 million	
Su-27SK fighter	Fourth- generation fighter	Sukhoi Design Bureau, KnAAPO	1996	1998- 2003	95 Su 27SK assembly kits	Licensed assembly in Shenyang	Chinese designation of the fighter made under license is J-11. China did not exercise the option for another 105 kits.
Su-27UBK combat trainer	Fourth- generation fighter	IAPO	December 1999	2000- 2002	28	Up to $800 million	Eight units delivered in 2000, 10 in 2001, 10 in 2002. Value of the deliveries offset against Russian sovereign debt to China.

PRODUCT	GENERATION	SUPPLIER	CONTRACT SIGNED	DELIVER-IES	UNITS	VALUE	NOTES
Su-30MKK multirole fighter	Fourth+- generation	Sukhoi Design Bureau, KnAAPO	August 1999	2000-2001	38	$1.8 billion	Ten units delivered in 2000, 28 in 2001
Su-30MKK multirole fighter	Fourth+- generation	Sukhoi Design Bureau, KnAAPO	July 2001	2002-2003	38	$1.8 billion	First 19 units delivered in 2002, the rest in 2003
Il-76MD transport	Developed in the 1960s	Tashkent Aircraft Plant	1992	1993	10	n/a	
Il-78 aerial refueling tanker	Developed in the 1970s	Unknown, possibly Ukraine	1998	n/a	4	Presumably $100 million	
A-50E AWACS	Developed in the 1970s	TANTK Beriyev, NIIP Tikhomirov	2001		4	n/a	Still being negotiated; talks began after Israel refused to supply Phalcon radars
R-73E short-range air-to-air missile	Developed in the 1970s	Vympel Corporation, Duks plant, Tactical Missiles Corporation	1995	1996-2001	1,200	n/a	Order placed for a total of 3,720 units (SIPRI estimate)
R-27 medium-range air-to-air missile	Developed in the 1970s	Vympel Corporation, Tactical Missiles Corporation	1990s	1990s	n/a	n/a	The main supplier of the R-27 missiles to China is GAKhK Artem (Ukraine) – up to 1,500 missiles
RVV-AE medium-range air-to-air missile with active radar homing device	Developed in the 1980s	Vympel Corporation, Tactical Missiles Corporation	2000	2001-2007	900	n/a	Orders placed for a total of 1,500 units (Western reports)
Kh-25 family air-to-surface missile	Developed in the 1970s	Zvezda Design Bureau, Tactical Missiles Corporation	1990s	1990s	n/a	n/a	The specific type of missile supplied was likely the Kh-25ML

PRODUCT	GENERATION	SUPPLIER	CONTRACT SIGNED	DELIVER- IES	UNITS	VALUE	NOTES
Kh-29TE air-to-surface missile	Developed in the 1970s	Vympel Corporation, Leningrad Severny plant, Tactical Missiles Corporation	1990s	1990s	About 2,000	n/a	Western estimates
Kh-31A anti-ship missile	Developed in the 1980s	Zvezda Design Bureau, Tactical Missiles Corporation	n/a	2003-2007	225	n/a	SIPRI estimate
Kh-31P (KR-1) anti-radar missile	Developed in the 1980s	Zvezda Design Bureau, Tactical Missiles Corporation	1994	Since 1997	n/a	License	Assembled in China, key components supplied from Russia. SIPRI estimate: 270 missiles, but in actual fact probably more.
Kh-59ME air-to-surface missile	Developed in the 1980s	Raduga Design Bureau, Smolensk Aircraft Plant, Tactical Missiles Corporation	n/a	Since 2002	n/a	n/a	Total of 150 missiles (SIPRI estimate), but in actual fact probably more.
Kh-59MK anti-ship missile	Developed in the 1990s	Raduga Design Bureau, Smolensk Aircraft Plant, Tactical Missiles Corporation	n/a	2006-2007	n/a	n/a	
APR-3ME airborne anti-submarine missile	Developed in the 1980s	GNPP Region, Tactical Missiles Corporation	1990s	n/a	n/a	n/a	For the Ka-28 helicopter
KAB-1500 Kr guided airbomb	Developed in the 1970s	GNPP Region, Tactical Missiles Corporation	n/a	n/a, probably after 2000	n/a	n/a	
AL-31FN aircraft engine	Fourth-generation aircraft engine	MMPP Salyut	n/a	1997-2001	10	Presumably $30 million	Engine for J-10 fighter prototype
AL-31FN aircraft engine	Fourth-generation aircraft engine	MMPP Salyut	2001	2002-2004	54	Estimated $150 million	Engine for J-10 fighter

PRODUCT	GENERATION	SUPPLIER	CONTRACT SIGNED	DELIVER-IES	UNITS	VALUE	NOTES
AL-31FN aircraft engine	Fourth-generation aircraft engine	MMPP Salyut	2005	2005-2006	100	$300 million	Engine for J-10 fighter
AL-31FN aircraft engine	Fourth-generation aircraft engine	MMPP Salyut	2007	2008-2009	100	$300 million	Engine for J-10 fighter
AL-31FN aircraft engine	Fourth-generation aircraft engine	MMPP Salyut	2005	unknown	150	Over $550 million	Contract signed in December 2005
RD-93 aircraft engine	Fourth-generation aircraft engine	OAO Klimov, MMP Chernyshev	2005	n/a	100	Estimated $270 million	For FC-1 fighter
Il-76 MD transports	Developed in the 1960s	Ilyushin, Tashkent Chkalov Aircraft Company	2005	n/a	34	Estimated at just over $1 billion	Contract has not entered into force
Il-78MK aerial refueling tanker	Developed in the 1970s	Ilyushin, Tashkent Chkalov Aircraft Company	2005	n/a	4	Estimated at about $120 million	

Ships and naval weapons

PRODUCT	GENERATION	SUPPLIER	CONTRACT SIGNED	DELIVER-IES	UNITS	VALUE	NOTES
Project 877EK diesel-electric submarine	Third-generation	Rubin, Krasnoye Sormovo	n/a	1994, 1995	2	$400 million	
Project 636 diesel-electric sub	Third-generation	Rubin, Admiralty Shipyards	n/a	1996, 1998	2	Up to $500 million	
Project 636M diesel-electric sub	Upgraded third-generation	Rubin, Admiralty Shipyards, Krasnoye Sormovo, Sevmashpred-priyatiye	2002	2004-2006	8	$2 billion	All will be equipped with the Club S missile system. Five of them were built by Admiralty Shipyards, two by Sevmashpred-priyatiye, and one by Krasnoye Sormovo in Nizhniy Novgorod. One sub delivered in 2004, six in 2005 and one in 2006

PRODUCT	GENERATION	SUPPLIER	CONTRACT SIGNED	DELIVER-IES	UNITS	VALUE	NOTES
Project 956E fleet destroyer	Third-generation	Severnoye Design Bureau, Severnaya Shipyards	1997	1999, 2000	2	$603 million for ships without weapons. Estimated value of the entire contract is up to $800 million	Manufacturer used the existing hulls of the *Vazhny* destroyer (70% ready in 1992) and the *Vdumchivy* (44% ready). Both were laid down in Soviet times.
Project 956EM fleet destroyer	Upgraded third-generation	Severnoye Design Bureau, Severnaya Shipyards	2002	2005, 2006	2	$1.4 billion	First destroyer delivered in 2005, second in 2006
Su-30MK2 multirole fighter	Fourth+-Generation	Sukhoi Design Bureau, KnAAPO	2002	2004	24	Over $1 billion	Optimized for use with Kh-31A and Kh-59MK anti-ship missiles
S-300FM Rif-M shipborne SAM system	Developed in the 1980s	NPO Altair, Almaz-Antey concern	2002	n/a, Presumably delivered in 2004 or 2005	2	n/a	For two Project 051C fleet destroyers
Shtil-1 shipborne SAM system	Developed in the 1980s	NPO Altair, Almaz-Antey concern	Presumably 2002	Presumably 2003	4	n/a	For two Project 052B fleet destroyers
Shtil-1 shipborne SAM system (vertical launch)	Upgraded circa 2000	NPO Altair, Almaz-Antey concern	unknown	Since 2006	4	n/a	For four Project 054A frigates, more deliveries may be in the pipeline
Ka-28 carrier-based anti-submarine helicopter	Developed in the 1970s	OAO Kamov, Kumertau Aircraft Company	1993	1993	2	n/a	
Ka-28 carrier-based anti-submarine helicopter	Developed in the 1970s	OAO Kamov, Kumertau Aircraft Company	1998	2000	5	n/a	
Ka-27PS carrier-based search and rescue helicopter	Developed in the 1970s	OAO Kamov, Kumertau Aircraft Company	1998	2000	3	n/a	

PRODUCT	GENERATION	SUPPLIER	CONTRACT SIGNED	DELIVER-IES	UNITS	VALUE	NOTES
3M80E Moskit anti-ship missiles	Entered service with the Soviet Navy in 1985	MKB Raduga, AAK Progress	1998	2000	50	Presumably $1.5 million per missile	For Project 956E fleet destroyers
3M80 MBE Moskit anti-ship missiles	Developed in the 1980s	MKB Raduga, AAK Progress	2002	2005-2006	50	Presumably $1.5 million per missile	For Project 956EM fleet destroyers
3M54E anti-ship missiles	Developed in the 1980s	OKB Novator, Kalinin Machine-Building Plant	2003	2005-2007	90	n/a	For Project 636M submarines. Total of 150 missiles ordered (SIPRI estimate)
AK-176 76mm shipborne artillery system	Developed in the 1970s	TsNII Burevestnik	1990s	n/a	License	n/a	Production under license
Pod-solnukh coastal surface wave effect radar	Developed in the 1980s	NPK NIIDAR	Unknown	2004-2006	2	n/a	

Air defense systems

PRODUCT	GENERATION	SUPPLIER	CONTRACT SIGNED	DELIVER-IES	UNITS	VALUE	NOTES
S-300 PMU-1 long-range SAM system	Developed in the 1980s	Defense Systems company	n/a	Until 1999	8 batteries	Up to $600 million	
S-300 PMU-1 long-range SAM system	Developed in the 1980s	Air Defense Concern	2001	2003, 2004	4 batteries	$450 million	Value offset against Russian sovereign debt to China. Two batteries supplied in 2003, two more in 2004
S-300 PMU-2 long-range SAM system	Early 2000s upgrade	Air Defense Concern	2004	n/a, Presumably 2006-2008	8 batteries	$980 million	China is the first, and so far the only, customer for this system
Tor-M1 short-range SAM system	Developed in the 1980s	Antey Concern, Izhevsk Kupol electro-mechanical plant	n/a	1996 and 1999	27 SAM systems	Listed price of one system: $27 million	Part of the value offset against Russian sovereign debt to China

PRODUCT	GENERATION	SUPPLIER	CONTRACT SIGNED	DELIVER-IES	UNITS	VALUE	NOTES
Ground weapons							
2K25M Krasnopol-M guided projectile	Upgrade of 1980s Soviet technology	Instrument Design Bureau, Izhmash	n/a	1999-2000	1,000		
2K25M Krasnopol-M guided projectile	Upgrade of 1980s Soviet technology	Instrument Design Bureau	1997	Ongoing	n/a	n/a	Production under license
Mi-17 transport helicopter	1960s-1970s	Kazan Helicopter Plant, Ulan-Ude Aircraft Plant	Series of contracts starting from 1990s	1990-2005	Over 300 units	Up to $1 billion	Some of the helicopters delivered to civilian customers
Bakhcha-U fighting compart-ment for armored vehicles	1990s	Tula Instrument Design Bureau	Probably 1995 or1996	1997	License	About $70 million	Production under license
9K116-3 Basnya 100mm round with 9M117M Kan missile	Developed in the 1980s	Tula Instrument Design Bureau	Probably 1995 or 1996	1997	License	unknown	Production under license
9K116-1 Bastion 115 round with 9M117 M3 Arkan guided missile	Developed in the 1980s	Tula Instrument Design Bureau	unknown	unknown	License	unknown	Production under license
9K119M Refleks-M 125mm round with 9M119M Invar guided missile	Developed in the 1980s	Tula Instrument Design Bureau	unknown	unknown	License	unknown	Production under license
RPO-A Shmel man-portable infantry flame-thrower	Developed in the 1970s	Tula Instrument Design Bureau	unknown	unknown	License	unknown	Production under license
2S23 Nona-SVK self-propelled 120mm mortar	Early 1980s	TsNII Precision Machine-building, Motovilikha Plants	Unknown	1997-1998	100	unknown	

PRODUCT	GENERATION	SUPPLIER	CONTRACT SIGNED	DELIVER-IES	UNITS	VALUE	NOTES
2S23 Nona-SVK self-propelled 120mm mortar	Early 1980s	TsNII Precision Machine-building, Motovilikha Plants	unknown	Since 2005	License	unknown	Production under license
9K58 Smerch MLR system mounted on 9A52 chassis	Early 1980s	GNPP Splav, Motovilikha Plants	unknown	unknown	License	unknown	Production under license
Fighting compart-ment and tipping unit of 2S19M1 Msta-S self-propelled mortar	Early 1980s	Uraltransmash, Tula Instrument Design Bureau, Plant No. 9 Design Bureau	unknown	Since 2005	License	unknown	Production under license

Source: Table compiled by CAST

Russian-Chinese military-technical cooperation from 1992 to 2007: three periods

There have been three distinct periods in Russian arms exports to China since the fall of the Soviet Union.

During the *first period,* which lasted from 1992 to 1999, China was buying export versions of standard Soviet weapons systems. The same systems, which represented 1970s to mid-1980s technology, were used by the Soviet Navy and Air Force. As with every other export version of Soviet weaponry, these systems were not exactly the latest and greatest products available. That was especially true of the Su-27SK fighter; its airframe was designed in the late 1970s, but its fire control systems (especially the N001 radar) represented 1960s technology. The same applies to the Project 956E (Sovremenny class) destroyers, whose final engineering designs were completed back in 1977. During this first period, the PLA made little use of the normal market practice of ordering weapons systems built to individual and specialized designs to fit the customer's exact requirements.

During the *second period,* which lasted from 1999 to 2004, China started to buy aerospace and naval weapons systems designed to individual specifications stipulated by the PLA. Those systems represented somewhat more advanced technology, but on the whole there was very little progress in that area compared to the previous period. The Chinese Air Force and Navy showed preference for fairly modest upgrades of standard-issue systems; reliability and good service record were far more important to them than risky technological innovation. The most notable contracts during the second period included the Su-30MKK multirole fighters and the upgraded Project 956EM destroyers. During this same period, China also became the launch customer for Russia's latest S-300MPU-2 SAM air defense systems.

Since the start of the *third period* in 2004, China has nearly stopped buying finished combat systems. The focus has shifted to combat support systems (such as transports and aerial refueling aircraft) and high-tech components used on domestic Chinese designs, i.e., engines, radars, homing heads for various missiles, etc. In addition, China has continued to place orders for airborne weapons, as well as transport and assault-landing helicopters. On the whole, however, Russian arms exports to China have fallen sharply since 2004. Measured by the value of new contracts signed, in 2006-2007 China was overtaken as Russia's biggest defense customer not only by India (which has always purchased large quantities of Russian weaponry), but also by two big new customers, Algeria and Venezuela. In 2009-2010, China was also overtaken by Vietnam. However, measured by the value of weapons deliveries under existing contracts, China was still first in 2004-2005, and first or second in 2006 and 2007.

Distinguishing features of Russian-Chinese military-technical cooperation

There are several key features that distinguish Chinese contracts for Russian weapons:

1. *Large volume.* By modern standards, China is buying weapons in very large batches. For example, in 1999-2003 it purchased 76 Su-30MKK and 24 Su-30MK2 fighter jets, worth up to $4.6 billion. That is comparable to the largest arms contracts being announced by such leaders in arms imports as Saudi Arabia and the United Arab Emirates. The contract for production under Russian license of 200 Su-27SK fighters is worth up to $2.2 billion. The order for eight Russian diesel-electric subs, which China purchased all in one go, was largely unprecedented by international standards.

2. *Relatively dated technology.* Unlike India, whose weapons technology policy is extremely bold and risky, China has been buying tried and tested weapons in an effort to minimize technological risks. That, however, may not necessarily be China's own preference; Russia itself continues to restrict the transfer of its latest defense technologies to the Chinese.

3. *No foreign components* to be integrated into Russian weapons systems ordered by China.

4. *Tight deadlines.* Unlike the long-term Indian contracts for Russian aircraft, Chinese contracts stipulate fairly tight deadlines. To illustrate, the 32 Su-30MKI fighters purchased by India were delivered in batches of 10-12 units over the course of three years. The Indian Su-30MKI licensed production program will take 12 years to complete. In contrast, deliveries on the 1999 Chinese contract for 38 Su-30MKK fighters were completed as early as 2001; Russia was delivering from 10 to 28 units each year. The second contract, signed in 2001, took only two years to complete, with 19 aircraft delivered in the first year and another 19 in the second. All deliveries on the Chinese contract for Su-30MK2 fighters were completed in one year, in 2004. On the one hand, such tight deadlines on large contracts keep the Russian supplier, Komsomolsk-on-Amur Aerospace Company, very busy. On the other hand, however, they mean that the company is unable to formulate a long-term financial and corporate strategy.

5. *Good commercial terms.* The Russian defense industry has enjoyed working with Chinese customers, especially in 1999-2003, for two main reasons. First, the Chinese buy in large batches. And second, they do not make any unreasonable demands regarding the modernization of the basic product range offered by Russian defense contractors. The practice of payment in kind, with shoddy Chinese consumer goods being supplied in exchange for Russian weapons, ended back in the mid-1990s. Now, in most cases Beijing pays in hard currency, or sometimes the value of the contract is offset against Russian sovereign debt to China.

6. *Military-political risks.* The risks of selling weapons to China are quite obvious, compared to customers such as India. The truth is that in terms of the sheer amount of resources available to Russia and China, the gap between the two countries is widening. This applies to military matters as well. The Russian military establishment is therefore quite wary of selling

large amounts of weapons to China, and continues to restrict the level of defense technology that Russia is prepared to supply to Beijing.

7. International repercussions. Another indirect consequence of Russian arms exports to China is that these exports hasten the process of China becoming America's main military rival. Obviously, it is not just Russia that is worried by China's military ascent—the US has similar concerns. And, truth be told, Moscow would actually benefit from Washington shifting the focus of its military planning from Russia to China.

2.3 Cooperation with Other Exporters

Although Russia remains the main arms supplier to China, there are other players on the Chinese defense market as well. Israel is the second most important player after Russia—not necessarily in absolute terms, but certainly as a source of defense technologies. Despite the European arms embargo, some European countries also supply some technologies and hardware that can be classed as military.

Israel

Defense industry cooperation between Israel and China began back in the late 1970s or early 1980s, even before the two countries formally established diplomatic relations. Initially this cooperation was welcomed by Washington, which tried to support China's efforts at modernizing its defense industry and army in order to ramp up military and economic pressure on the Soviet Union. Israel, meanwhile, tried to extract political dividends in addition to commercial gain by pressuring China to limit its arms sales to Arab countries. Chinese-Israeli defense industry cooperation peaked in the early 1990s, after Europe imposed its arms embargo on Beijing. For a period of time, Israel became China's second most important source of defense technology after Russia. During that period, the two countries worked closely on the J-10 fighter project, among other things. In 1999, China accounted for 20% of Israel's $1.2 billion arms exports (i.e., $240 million).

However, in the mid-1990s Washington completely reassessed its stance on

Israeli-Chinese defense industry cooperation. America's confrontation with the former Soviet Union had become a thing of the past, whereas China's ascent—in particular, its growing military capability—was causing growing concern in Washington. The US began to apply mounting pressure on Israel to end its contacts with Beijing. Tensions between Washington and Tel Aviv reached their peak when Israel tried to sell its Phalcon radar system to China. It was eventually forced to yield to US pressure and cancel the contract, paying the Chinese $350 million in penalties. It appears that sometime in 2002-2003, Washington finally got Israel to radically curtail its cooperation with China. At any rate, in 2003 Israel Aerospace Industries (IAI), which was at the forefront of that cooperation, drastically reduced the size of its office in Beijing. Israeli sales to China now appear to be limited mostly to equipment used by anti-terrorism units and the police.

One major Chinese project in which Israeli involvement is felt to this day is the J-10, a light fighter jet based on the Israeli IAI Lavi design for a highly maneuverable attack aircraft. Cooperation on this project began in the mid-1980s and was stepped up after the Lavi program was cancelled in Israel. China bought many of the design and engineering documents left from the program and used them to start developing its own aircraft. Later on, Chinese engineers were forced to make significant changes to the original Israeli specifications, because the only engine available for the new aircraft was the Russian AL-31FN. In addition to the basic Lavi designs, Israel also sold to China several sets of control systems for the plane, and may have also transferred the underlying technology. Israel also appears to have taken part in the development of radioelectronic systems for other Chinese aircraft, including the J-8-II fighter.

For China, Israel is also an important supplier of short-range air-to-air missiles and the underlying manufacturing technology. In 1982, the two countries signed an agreement on joint production of the Python 3 missile in China (Chinese designation PL-8). Production began sometime in 1986-1988. The PL-8 is now the main weapon system used on domestic Chinese fighters, including the J-10. According to some American sources, Israel sold China several units of the Python 4, an advanced short-range air-to-air missile. It may have even sold the underlying technology as well. At an airshow in Zhuhai in 2002, China's AVIC-1 aerospace concern showed a video clip of a J-10 firing a missile that looked very similar to the Python 4.

Another important area of Israeli-Chinese cooperation is unmanned aerial vehicles. In 2000, Israel supplied to China an undisclosed number of the IAI Harpy attack UAVs, which are used to disable radars. Cooperation between the two countries on ground weapons appears to have been limited to just two projects, but both were fairly important to China. The first was the transfer of Israeli technology for producing tank ammunition with depleted uranium core. The second was the use of Israeli technologies in the Chinese HJ-9 anti-tank missile system. The Americans believe the design of this weapon was "influenced" by the Israeli MAPATS system and made use of several Israeli technologies.

Israeli arms and defense technology exports to China have not been large in dollar terms. But they have been very important to China as a means of plugging various gaps in imports from its main supplier, i.e., Russia. The end of defense industry cooperation with Israel has significantly slowed China's progress in several important areas, especially its projects to develop long-range radar systems.

Europe: Before the arms embargo

Defense industry cooperation between China and Europe reached its peak in the late 1970s and early 1980s. There were three main areas of cooperation: helicopters, naval weapons and air defense systems.

Helicopters. France and Italy have probably played the biggest role in the development of China's helicopter industry. Three types of French helicopters are now produced in China, either under license or as pirated copies. Back in 1980, France and China signed an agreement on joint production of the Aerospatiale AS365N Dauphin 2 medium helicopter in Harbin. The Chinese version of the helicopter was designated as the Z-9. By 1992, Harbin Aircraft Manufacturing Corporation (which has since changed its name to HAIG) completed the transition from merely assembling the helicopter from French components to producing a localized version from locally made components. The Z-9 is now the most widely used helicopter in China, both in the PLA and in the civilian sector. In 1989, China also launched small-series production of the French Aerospatiale SA321 Super Frelon heavy transport helicopter, designated locally as the Z-8. Finally, in 1997 Changhe Aircraft Industry Corporation (CHAIG) launched production of a pirated copy of the Aerospatiale AS350 Ecureuil light helicopter, designated locally as the Z-11. The helicopter is used for training PLA pilots; CHAIG is also actively marketing it to civilian customers.

As part of an agreement with Eurocopter in 2005, HAIG launched production of the Eurocopter EC120 light helicopter (designated locally as the HC120) in Harbin. The two companies are also jointly developing the EC175 (Z-15) medium helicopter. There have also been reports that Eurocopter and AgustaWestland have been involved in the development of the Z-10 attack helicopter. Finally, CHAIG has signed an agreement to assemble the AgustaWestland AW109 (local designation CA109) light helicopter at its plant in Changhe.

Shipborne weapons systems. In addition to the Chinese helicopter industry, France has left its mark on many of China's naval weapons programs. This includes the Creusot-Loire T100C 100mm shipborne artillery system. China bought only two units. One of them was installed on the Sipin frigate; the other was used for reverse engineering. At this time, several of the latest Chinese ships, including the Project 052B and 052C destroyers and Project 054 frigates, are equipped with a 100mm artillery system that appears to be a pirated copy of the T100C. China used a similar strategy with the Navale Crotale SAM system, first buying just enough units to equip two ships, then reverse engineering some of them and launching mass production of a pirated copy. Beijing also relied heavily on French technologies for the development of YJ-8 anti-ship missiles and sonar systems for submarines, including nuclear subs. Finally, the Chinese used the Dutch Goalkeeper 30mm air defense artillery system in the development of their own Type 730 system.

Short-range air defense systems. Cooperation with Europe had a very strong influence on Chinese programs to develop short-range air defense systems in the 1980s. The Chinese used the French Crotale SAM system as a prototype for their own HQ-7 (FM-80 and FM-90), which they are now making in large numbers and which will probably be their primary low- and medium-altitude air defense system for years to come. The success of the HQ-7 program appears to have been the main reason why the Chinese have stopped buying the Tor-M1 from Russia. Finally, in the 1990s China used the Italian Aspide air-to-air and SAM missile, for which it had acquired a license, to develop the HQ-64 (LY-60) short-range SAM air defense system, but the weapon does not appear to have entered service with the PLA.

The Chinese aerospace industry has worked with Italian and British companies to upgrade the J-7 fighter (a copy of the Soviet MiG-21), which makes up the core of the Chinese Air Force fleet. Some of the J-7 modifications were equipped with the Italian Grifo 7 radar; cooperation in this area with Italy continued even after Europe imposed its arms embargo on China.

Key Chinese weapons programs which rely on European technologies

PROGRAM	COUNTRY	PERIOD	CURRENT STATE
AS365N Dauphin 2 helicopter	France	1980s	Several versions being produced in China. This is the main Chinese military helicopter.
Crotale, Navale Crotale SAM systems	France	1980s	Produced in China. Main short- and medium-range SAM system serving with the Chinese Navy and Army.
AS350 Ecureuil helicopter	France	1990s	Pirated copy made in China
SA321 Super Frelon helicopter	France	1970s–1980s	Small-series production; Chinese design based on helicopters bought from France
SA342L Gazelle helicopter	France	1980s	Eight helicopters bought from France now in service with the PLA
Creusot-Loire T100C 100mm universal shipborne artillery system	France	1980s	Chinese design based on units bought from France; mass production ongoing. This is China's main universal shipborne artillery system used on new-generation ships (Project 052B and 052C destroyers, Project 054 frigates)
Submarine sonars	France	1980s–1990s	Used on Han class submarines and some of the Ming and Song class boats
Han-class submarine	France	1980s	French technical assistance to improve main propulsion unit
Z-10 helicopter	France, Italy	1990s to date	Development of a future helicopter in cooperation with Eurocopter, Agusta

PROGRAM	COUNTRY	PERIOD	CURRENT STATE
Alenia Aspide SAM system	Italy	1980s	Chinese LY-60 design based on units bought from Italy; did not reach mass production
NJ2405, NJ2406 off-road trucks	Italy	1990s	Light army trucks made by IVECO in Nanking
Grifo 7 radar	Italy	Late 1980s– early 1990s	The Italian radar is used on some J-7E fighters
Rolls-Royce Spey 202 engine	Britain	1970s to date	Used on JH-7 and JH-7A bombers. The Chinese seem to have acquired a license.
Searchwater radar	Britain	1990s	Used on six to eight Y-8J, the Chinese Navy Air Force's main patrol aircraft
GEC-Marconi Super Skyranger radar	Britain	1990s	Used on some J-7E fighters
Micro- and nano-satellites	Britain	1990s to date	Surrey Satellite Technology Ltd. is helping Chinese scientists from Tsinghua University in the development of small satellites. A number of satellites have been designed and manufactured; cooperation is ongoing.
Tiema XC2200 trucks	West Germany	1980s	Heavy army trucks, design based on the Mercedes-Benz 2060, possibly designed with German technical assistance
Song-class diesel-electric submarine	Germany	1990s	Technical assistance in designing the sub's diesel
Galileo satellite navigation system	EU	Ongoing	China is taking part in Europe's Galileo satellite navigation project

Source: Table compiled by CAST

Europe: After the embargo

In June 1989, the European Economic Community's Council of Foreign Ministers passed a resolution to impose an arms embargo on China. The decision had profound implications for China's defense industry policy. Beginning in the early 1990s, Beijing has been forced to rely on Russia as the main source of advanced weapons technologies. Nevertheless, cooperation between Europe and China in this area has not stopped completely. The resolution passed by Europe's foreign ministers did not contain a precise definition of the word "embargo"; interpretation of it was left to the national governments. Those governments that had important commercial interests in China in the late 1980s (especially France, Italy and Britain) adopted an extremely narrow definition, which applied only to combat systems. That left them enough leeway to continue supplying other military hardware and technologies, including airborne and ground radars, components for military satellites, range sensors, video surveillance and targeting systems, electronics, telecommunication and navigation equipment, instruments for testing small arms ammunition, explosives for military engineering applications, and much more. As a result, European supplies to China continued even after 1989, albeit on a much smaller scale. For example, in 1996 Britain supplied six Searchwater radars worth $66 million; the radars are used on the Y-8J, the Chinese Navy Air Force's main patrol aircraft. Italy has sold Grifo 7 radars for the J-7E fighter under a contract signed in 1993. In addition, in the early 1990s European suppliers continued deliveries on contracts signed prior to the sanctions. We estimate that Europe has been selling China $200 million to $300 million worth of military and dual-use equipment every year.

2.4 Latest Trends on the Chinese Market

Clear new trends have emerged since 2004, suggesting a radical shift in the Chinese defense procurement policy. Namely, China is:

- Reducing arms imports from Russia;
- Shifting its focus from buying entire weapons platforms to importing critical components such as engines and avionics, as well as airborne and shipborne weapons;
- Transitioning from buying weapons to importing technologies;
- Maintaining hope that the European arms embargo will be lifted;
- Stepping up procurement for internal security forces, which was especially notable ahead of the 2008 Beijing Olympics;
- Moving closer to becoming a net exporter of weapons once again, and potentially to emerging as a major competitor to Russian arms exporters.

The most obvious of these trends is the sharp fall in new orders placed with Russia by the Chinese. The first signal came when China chose not to exercise its option for an additional 105 Su-27SK kits for assembly in Shenyang after receiving 95 such kits in 1998-2004. Now that it has eliminated teething problems with the mass production of the J-11 (the Chinese designation of the Su-27SK produced under license), China has opted instead to develop its own clone of the fighter, known as the J-11B. Russia's hopes that China will order a second regiment-sized batch of Su-30MK2 naval fighters have also been dashed. The first 24 of these fighters, which are configured to attack ships, were delivered in 2004, and China was expected to buy another 24. The last known Chinese contract for Russian aircraft came in 2005, when Beijing placed an order for 34 Il-76MD transports plus four Il-78MK aerial refueling tankers. But a year into the contract it became clear that the supplier, the Tashkent Aircraft Plant, would be unable to deliver. The company had run out of existing airframes and components left from Soviet times, and was hit by the falling exchange rate of the US dollar, in which the contract was denominated. As a result, Russia did not go ahead with that contract and it fell through.

The last known Chinese aircraft contract on which deliveries have now been successfully completed dates back to 2003. Since then, China has been placing orders mostly for upgrades of previously supplied aircraft, as well as for new airborne weapons, aircraft engines, and shipborne weapons and air defense systems. The only exception is transport and assault-landing helicopters; China continues to buy the Mi-17.

As a result, there has been a steep fall in the value of new Chinese contracts. According to Rosoboronexport chief executive Sergey Chemezov, in 2006 China placed a mere $200 million worth of orders for Russian weaponry. The figure rises if we take into account contracts signed with some other Russian weapons exporters, such as Sukhoi, which is allowed to sell directly to foreign customers, bypassing Rosoboronexport. But the overall value of new Chinese orders is still below $500 million. We estimate that in 2007, it rose slightly to $700 million to $800 million. Nevertheless, if these estimates are correct, in 2006-2007 China ranked fourth or even lower on the list of Russia's biggest defense customers, after India, Algeria and Venezuela.

The value of Russian weapons deliveries to China remained fairly high—at about $1.5 billion to $2.5 billion a year—in 2004-2007, thanks mainly to contracts signed in previous years. During that period, Russia delivered eight Project 636M submarines and two Project 956EM destroyers, plus various airborne weapons and naval weapons systems. By value of deliveries, China still remained the top Russian defense customer, with the exception of 2004 and 2007, when India claimed the top spot.

The most likely explanation for the steep fall in new Chinese defense contracts is that the PLA is no longer prepared to settle for buying late 1980s or early 1990s technology. Besides, such technology can already be sourced from defense contractors within China itself. In 2005-2006, China received the last batch of very conservatively upgraded fourth-generation aircraft and naval systems. Now it is interested only in much more advanced weaponry. In addition, it insists on buying very small batches, preferring to import technology rather than finished products. And unlike India, China is pursuing a policy of "defense industry nationalism," showing little enthusiasm for joint projects to develop, produce and market next-generation weapons.

As for the Russian defense industry, it managed to secure large orders from Algeria, Venezuela and some Middle Eastern states in 2005-2006. Russian Ministry of Defense contracts are also on the rise, so Russian defense companies no longer depend on Chinese contracts for their survival. Meanwhile, the Russian military establishment's opposition to transferring the latest weapons technologies to China is gaining traction as China's military might and defense industry capability continue to grow. The mutual interest that had fueled Russian-Chinese arms trade over the past 15 years has therefore all but fizzled out. Although opportunities for cooperation still remain in individual segments of the defense market, the era of huge Chinese contracts for Russian weapons seems to be well and truly over.

European arms embargo

Another apparent reason for China's waning interest in Russian weapons is its hope that Europe will soon lift the arms embargo. Starting in 2003-2004, some European countries, especially France and Italy, have been campaigning to end restrictions on arms exports to China. While America sees China as an emerging rival superpower, and while Russia frets about this new superpower's close proximity, Europe does not see China's growing military might or defense industry capability as a security threat. On the contrary, China's rise opens up attractive possibilities for the European defense industry. It would not be a stretch to assume that if Europe were to lift the arms embargo its defense contractors would be far less reluctant than their Russian counterparts to supply the most advanced weaponry to Beijing. After all, it is very difficult for Russia to ignore the fact that it is increasingly falling behind China both economically and militarily.

France, in addition to campaigning against the arms embargo, is already marketing some of its weapons systems in China. Dassault has held a presentation of its Rafale fighter (or possibly its carrier-based version, the Rafale M) in the PRC. Chinese specialists have also expressed interest in having a much closer look at the Snecma M88 turbofan engine. French and Italian lobbying for the resumption of large-scale defense industry cooperation with China may have given Beijing the hope that it will soon have another source of advanced defense technologies. But as of spring 2008 the embargo is still in place. Any attempts to lift it are probably being blocked by Washington's allies in the EU. Washington, meanwhile, is not driven solely by military-political considerations. From the commercial point of view it does not want the French defense industry to benefit from Chinese contracts. French companies are American defense contractors' main rivals in many markets, including even the United States itself.

Prioritization of technology transfers

The Chinese economy has been growing at an astonishing pace in recent years, and the defense industry is no exception. Its achievements have been especially obvious in military shipbuilding, which draws on the strength of China's commercial shipbuilding sector. In the aerospace industry, progress has been less spectacular but still very respectable. All the Russian specialists who have had a chance to visit Chinese aerospace facilities remark on how well they are equipped. China has become the leading buyer of advanced machine tools, indicating the scale of the Chinese technology upgrade programs in machine-building. The country has become the world's workshop in civilian industries. That has translated into growing defense industry capability as well. Beijing is no longer interested in buying finished products; it wants the underlying technology. More than half of the ongoing Russian-Chinese military and technical cooperation projects involve technology development or transfer. This focus on technologies rather than finished products is opening up excellent opportunities for European and Israeli companies, which have to take into account America's attitude toward arms trade with China.

Growing interest in equipment used for domestic security purposes

In the run-up to the 2008 Beijing Olympics, there was a steep increase in Chinese imports of equipment used by the police and domestic security agencies. It appears that most of those imports were from Western countries and Israel.

Prospects for China regaining its status as a net exporter of weapons

In the second half of the past decade, China developed and brought to mass production status many weapons systems that have excellent export potential. The country is well on its way to returning to the global defense market as a leading supplier. Its most competitive products are the FC-1 (JF-17) and J-10A light fighters. Both represent fourth-generation technology and can compete head to head with the Russian MiG-29 (including the latest and most advanced MiG-29SMT version). On the Pakistani market, they can also give the American Lockheed Martin F-16 a run for its money. The main competitive advantage of the two Chinese aircraft is that they offer an acceptable level of technology at affordable prices. The FC-1 was designed with the export market in mind right from the start. Pakistan, which was involved in its development, became the launch customer. According to the most conservative estimates, the Pakistani Air Force's requirement for the FC-1 is at least 250 units, although the actual contracts will probably be much less ambitious. Apart from Pakistan, the FC-1 can be successfully marketed to poorer countries, where the most popular models now include the F-7 (which represents the previous generation of Chinese technology), the Russian MiG-29 and the Israeli IAI Kfir. These markets include Ecuador, Bolivia and Peru in Latin America, Burma in Southeast Asia, Bangladesh and Sri Lanka in South Asia, and Azerbaijan in the CIS. Another potential market is the whole of sub-Saharan Africa, with Zimbabwe, Tanzania, the Democratic Republic of the Congo and Eritrea being the most likely customers. The heavier and more expensive J-10 has good prospects on the fairly large markets of Venezuela, Iran and possibly Nigeria. The Pakistani Air Force is also showing interest in the J-10; there have been reports that 36 to 40 of them could be purchased in addition to the FC-1. Finally, China's own Air Force will be a very big customer; large domestic contracts will serve as yet another competitive advantage of the J-10. The main weakness of the two Chinese fighters is that China remains dependent on imports of the Russian Al-31FN and RD-33 engines used on the J-10 and FC-1, respectively. But the Chinese defense industry is quite likely to produce domestic replacements with an acceptable lifespan in the not too distant future.

The Chinese military shipbuilding industry may also win some foreign contracts. Chinese ships may find a niche in the lower price segment of the markets where China already has a strong presence. It is not unrealistic to find buyers for Chinese-built hulls, which can then be outfitted with Western weapons and electronics at the customer's own shipyards. The Chinese shipbuilding industry's greatest success so far has been the 2005 Pakistani contract for four Project F-22P frigates, to be delivered by 2013. Finally, China remains fairly competitive in the ground weapons segment of the defense market.

On the whole, China has bright prospects in two types of markets. The first is poor countries in Africa, South Asia, Southeast Asia and Latin America, which need simple and affordable weapons. The Chinese industry is very cost-competitive as it is, and the Chinese government can subsidize some deliveries, if need be. The second type of market is countries that pursue distinctly anti-Western or anti-American policies. These countries are either denied access to European and American defense technology or buy from non-Western suppliers as a matter of principle. These include Venezuela, Iran and possibly Syria. In the first type of markets, China will compete head to head with suppliers from the CIS and Central and Eastern Europe, which specialize in weaponry they inherited from the former Soviet Union and the Warsaw Pact countries. In the second type of market, China's main competitor will be Russia. Selling weapons to anti-Western regimes requires political will and resilience to economic and political pressure from the US, Europe and Israel. Beijing appears less sensitive to such pressures than Moscow, so China's guarantees regarding the reliability of supplies will look more persuasive than Russia's. As its level of technology continues to improve, the Chinese defense industry will become a serious competitor to Russia's, not only in China itself but in other markets as well.

Finally, there is every reason to expect stronger Chinese presence in countries that pursue a balanced defense technology procurement policy and try to diversify the sources of their weapons imports. These include Pakistan, which is a de facto military-political ally of China, as well as other traditional Chinese defense customers such as Egypt and Thailand.

A Chinese warship fires missiles during the "Peace Mission 2005" China-Russia joint military exercise off China's Shandong Peninsula, August 23, 2005. The exercise involved nearly 10,000 military personnel from the two countries. *REUTERS/China Newsphoto*

The People's Liberation Army Navy guided missile destroyer *Shenzhen* (DDG 167) sails into Apra Harbor, Guam, on October 22, 2003—the People's Republic of China's first ever port call to Guam. *US Navy photo by Photographer's Mate 2nd Class Christopher S. Borgren II/ Released*

The Chinese aircraft carrier *Varyag* nears completion. *Patriots Point Naval and Maritime Museum*

Chinese amphibious armored vehicles take part in a China-Russia joint military exercise in eastern China's Shandong peninsula, August 24, 2005. *REUTERS/China Newsphoto*

Chinese Chengdu J-10 fighters take off, November 17, 2010. *Peng Chen*

A Chinese Su-27 fighter flies over Anshan Airfield, China, during a visit by the Chairman of the Joint Chiefs of Staff, Marine Gen. Peter Pace, March 24, 2007. *US Department of Defense photo by Staff Sgt. D. Myles Cullen, USAF/Released*

A Chinese F-8 flies near a US Navy aircraft, April 1, 2009. *US Department of Defense/Released*

The People's Liberation Army Air Force J-10 fighter jet aerobatic team fly in formation during a rehearsal for the upcoming 60th anniversary of the founding of the PLA Air Force near an airport on the outskirts of Beijing, October 31, 2009. *REUTERS/Joe Chan*

A People's Liberation Army Air Force Jian-10 (Chengdu J-10) fighter jet flies at Yangcun Air Force base on the outskirts of Tianjin municipality, April 13, 2010. *REUTERS/Petar Kujundzic*

Chinese Marines stand at attention on a warship with a Changhe Z-8 helicopter while taking part in an international fleet review to celebrate the 60th anniversary of the founding of the People's Liberation Army Navy in Qingdao, Shandong province, April 23, 2009. *REUTERS/Guang Niu*

A People's Liberation Army Air Force Z-8KH helicopter, May 1, 2010. *Alancrh*

An unmanned aerial vehicle is seen during a parade to mark the 60th anniversary of the founding of the People's Republic of China in Beijing, October 1, 2009. *REUTERS/David Gray*

A ZBD-05 amphibious infantry fighting vehicle takes part in rehearsals in Beijing on September 6, 2009 for the upcoming parade to celebrate the 60th anniversary of the founding of the People's Republic of China. *Dan Cheng*

A Chinese anti-aircraft missile system and launcher, May 8, 2010. *Xi Zhang/ Dreamstime*

A Chinese S-300 (HQ-9) launcher in China's 60th anniversary parade, October 1, 2009. *Jian Kang*

Surface-to-surface missiles are displayed in a parade to celebrate the 60th anniversary of the founding of the People's Republic of China, October 1, 2009. *REUTERS/Jason Lee*

A soldier of the Chinese People's Liberation Army checks a PLA ZTZ-99 tank during the "Peace Mission 2010" exercises at the Matybulak military range in southern Kazakhstan, September 24, 2010. *REUTERS/Shamil Zhumatov*

Chinese arms on display at the "Our Troops Toward the Sky" exhibition at the Bejing Military Museum, August 2007.

Top: A Type 95 SPAAG vehicle

Middle: A KS-1 surface-to-air missile mobile launcher

Bottom: A PLZ05 155mm self-propelled howitzer

All photos by Max Smith

Chapter 3
Chinese Arms Exports

3.1 From 1949 to 1992

There have been three distinct periods in Chinese arms exports between the foundation of the People's Republic of China and the end of the cold war:

- Ideological (1950s-1960s)
- Geopolitical (early 1970s)
- Commercial, with an export boom in the 1980s during the Iran-Iraq War

Communist China began exporting weapons in the 1950s, when shortly after the foundation of the PRC, the government in Beijing decided to provide military assistance to Communist forces in French Indo-China (Vietnam) and to North Korea during the Korean War. In both cases, China supplied Japanese and US weapons seized as spoils of war in 1937-1949, or acted as an intermediary for Soviet arms suppliers. The country's own defense industry was in an embryonic state at that time.

With Soviet help, that industry made significant progress in the late 1950s, and in 1959 China began to supply its own weapons to foreign countries. The main recipients were China's allies, such as Albania, North Vietnam and North Korea, as well as various "national liberation" movements supported by Beijing. In the 1960s, China began supplying its weapons to newly independent African nations as part of its efforts to win greater influence among the developing countries. The massive military support provided by China to North Vietnam and the Vietcong was a major factor in the Communists' victory in Indo-China.

It is safe to say that in the early period, Chinese arms exports were led by ideological considerations; commercial motives were either totally absent or played only a marginal role.

In the late 1960s China established close ties with Pakistan, which quickly became the biggest importer of Chinese weapons (and remains so to this day), including heavy and advanced weaponry. Massive Chinese arms exports to Pakistan became the foundation of a strategic military-political alliance between the two countries. The beginning of large-scale cooperation with Islamic, capitalist and pro-Western Pakistan marked the end of the period in which Communist China's choice of military partners was informed solely by ideological motives. It signaled the arrival of Beijing's Realpolitik strategy in the early 1970s (a visit by Richard Nixon, etc.), which prioritized pragmatic political and military considerations over ideology. But pragmatism quickly degenerated into a complete lack of principles, as other Communist governments saw it. During the same decade, China established military ties with a number of right-wing and anti-Communist regimes in the "third world" (such as Zaire and Egypt). Amid growing confrontation with the Soviet Union, it even began to provide assistance to anti-Communist insurgencies in Angola and Afghanistan. It also made its first attempts to import Western military technology during this period. But arms supplies to ideologically close countries and movements, such as North Korea, the Khmer Rouge (including their years in power in Cambodia) and Albania continued to play an important role.

The Iran-Iraq War (1980-1988) was a major stimulus for Chinese arms exports. Thanks to the absence of any direct Chinese interests in the region, as well as to China's ability to offer large quantities of affordable weapons, the country became one of the leading suppliers of weapons to both Iran and Iraq. During that period China emerged as a big net arms exporter, and one of the world's leading arms suppliers. Its strained relations with the West (and to some extent with the Soviet Union as well) helped Beijing become a privileged partner of Tehran. As a result, the Islamic Republic became China's second biggest long-term defense customer after Pakistan. This period, when China's export policy was led by motives of profit, can be described as the "commercial stage."

3.2 After 1992

China's arms exports since the breakup of the Soviet Union can likewise be divided into three periods.

During the *first period*, which lasted until 2000, exports fell sharply to about $800 million a year, and most of them were destined for Pakistan or Iran.

During the *second period*, from 2001 to 2005, China managed to break into the Kuwaiti market and increased its presence in Latin America and Africa. Annual exports rose to about $1 billion.

During the *third period*, China won such new customers as Saudi Arabia, Morocco, Venezuela, Ecuador, Peru and Indonesia. Exports rose sharply, even reaching $1.8 billion some years.

The breakup of the Soviet Union opened up what seemed like excellent opportunities for Chinese arms exports. Theoretically Beijing was in a good position to replace Moscow as the main supplier of weapons to traditional Soviet allies in the Middle East, Africa, Asia and Latin America.

China made some efforts to win these new markets, but it was hampered by the woeful technological level of its national defense industry. By the early 1990s, all the weaponry China had been receiving (and copying) directly from the Soviet Union until the mid-1960s, or by way of third countries until the 1970s, had become hopelessly obsolete. Almost all the upgrade options for that weaponry had been exhausted. China itself lacked the ability to develop a new generation of weapons, and the imports of Western military technology were shut off abruptly following the Tiananmen Square protests in 1989 and the ensuing arms embargo.

That is why China largely failed to capitalize on all the new opportunities to bolster its arms exports in the 1990s. Instead, it became a very large importer of modern Russian weapons and defense technologies while it restructured and modernized its own defense industry. This marked the second period in the PRC's history since the 1950s and 1960s during which the country became a large net importer of arms and defense technology.

Pakistan and Iran remained the biggest Chinese defense customers throughout the 1990s. Both countries began to partner with the Chinese as co-financiers of defense R&D projects. Starting in the late 1980s, the list of the biggest buyers of Chinese weapons also included Burma and Thailand. Technologically the weaponry supplied to all these countries remained extremely dated. Also during this period, China was actively exporting missile technologies (medium- and long-range missiles), because none of the other leading players were present in that segment of the market for political reasons.

We estimate that China sold about $7 billion worth of arms to foreign customers from 1992 to 2000. That is not a very high figure, and it puts China in the second-tier category of global arms suppliers. Nevertheless, a few of the Chinese systems sold during that period were already able to compete with Soviet- and Russian-made weaponry. This applied to segments such as small arms, infantry weapons and artillery systems, with lower prices being the main competitive factor.

The first decade of the 2000s marked the rollout of advanced technologies licensed or pirated from Russia and the CIS across the Chinese defense industry. China began manufacturing many weapons systems under Russian license, while at the same time developing a broad range of domestic new-generation systems. Its weapons R&D projects relied either on copying foreign technology or incorporating imported components into its own designs. The process is lengthy and complex, and in many ways it is still ongoing. Chinese technology has yet to catch up with the world leaders in several crucial areas, such as engines, targeting systems, radars, etc.

In 2001-2005, China continued to supply obsolete weaponry whose origins dated back to Soviet technology of the 1950s and 1960s. But defense industry modernization programs were starting to yield practical results. During that period, China sold some fairly advanced weaponry to Pakistan, such as C-801 and C-802 anti-ship missiles, man-portable SAM systems, K-8 jet trainers, PLZ-45 self-propelled howitzers, and Al Khalid tanks (Chinese Type 90 model). Pakistan's role as the leading buyer of Chinese weapons was reinforced in the early 2000s after Islamabad came under US sanctions over its nuclear weapons program. Iran also remained an important customer. China's growing global influence, which was based on its economic achievements, also enabled it to secure important contracts for artillery systems with

Kuwait, to break into the Latin American and Indonesian markets, and to strengthen its presence in Africa.

As a result, Chinese arms exports showed substantial growth in 2001-2005, during which the country sold an estimated $5 billion worth of weaponry. Much of those exports went to Pakistan. For example, Islamabad bought 57 F-7MG fighters—clearly for lack of a better alternative, since the aircraft were rather obsolete. Yet, with annual arms exports still in the $1 billion range, China remained a second-tier supplier.

However, the period of 2005-2010 brought rapid Chinese expansion on the global arms market. Thanks to its breakneck technological growth, China is now able to offer foreign customers a number of advanced new weapons systems. The country's progress has been especially impressive in the aerospace sector. Its two fourth-generation fighter programs, the FC-1 (developed jointly with Pakistan) and the J-10, have reached mass production stage. Both aircraft are now being marketed abroad, especially the FC-1. Pakistan was the launch customer for the FC-1 and for the export version of the J-10, reaffirming its status as China's most important defense industry partner. China has also been able to offer Pakistan its new Project F-22P frigates and the Z-9C shipborne helicopters. It is energetically marketing its new missile systems (anti-ship, anti-tank and man-portable SAM), radars, transport aircraft (MA60), helicopters (Z-9 and Z-11), new versions of the Type 90 tank, a new generation of light armored vehicles (which has found a market in Africa), self-propelled and towed artillery, trucks and ships.

In recent years, China has broken into several important new markets such as Saudi Arabia, Morocco, Venezuela, Ecuador, Peru, Mexico, Nigeria, Kenya and Indonesia. Exports in the five years prior to 2010 rose sharply, to an aggregate $9 billion. Pakistan still remains the key customer, but there is no denying the much broader product range and geographical distribution of Chinese arms exports.

Especially noteworthy is the increase in Chinese arms sales to Latin America and Africa, which Beijing clearly views as priority markets for expansion. Chinese bids for weapons contracts in these countries often include a broader package of economic, trade, and R&D cooperation. It is easy to see the attraction of Chinese offers for the poorer countries in these two regions.

In the third world countries, China is trying to position itself as an alternative to the Western powers. That has clear implications for these countries' military ties with Russia. They can use China as a counterbalance not only to Western but to Russian influence as well. The CIS nations will find that strategy especially tempting.

Chinese presence on the global arms market in 1992-2010

STAGE	ANNUAL SALES IN DOLLARS	DESTINATIONS	WEAPONS
1992-2000	800 million	Pakistan, India	Small arms, artillery systems
2001-2005	1 billion	First sales to Kuwait, increased presence in Latin America and Africa	F-7MG fighter, K-8 trainer, PLZ-45 self-propelled artillery, Al Khalid main battle tank
2006-2010	1.8 billion	First sales to Saudi Arabia, Morocco, Venezuela, Ecuador, Peru and Indonesia	FC-1 and J-10 fighters, F-22P frigates, Z-9 and Z-11 helicopters

3.3 Biggest Importers of Chinese Weapons Since 1992

China does not release official statistics about its arms exports. It began submitting reports to the UN Register of Conventional Arms only in 2006, but these reports are clearly incomplete and not very informative; Beijing seems to treat them as a mere formality. Much of the information below about Chinese arms exports is therefore based on estimates, rather than hard official data.[99]

Traditionally, China's defense customers have been either Beijing's military-political allies (Pakistan and North Korea, as well as Albania and Romania in earlier decades) or pariah states that are denied access to Western weapons. Iran, Burma and Sudan fall into the latter category. Another important partner since the late 1970s has been Egypt. Pakistan, North Korea, Iran, Sudan, Burma and Egypt are in fact China's biggest defense customers in terms of dollars.

For all the progress the Chinese defense industry has made over the past decade, its key markets are still limited to developing countries in Asia, Africa, and in recent years Latin America. Affordable prices remain the key attraction of Chinese weaponry for the world's poorest nations.

But another new trend that has emerged in recent years is that some countries have started to buy from China for reasons of political diversification. That seems to have been one of the primary considerations for rich buyers such as Kuwait and Saudi Arabia, as well as for Turkey and Nigeria, which also have sizable defense budgets. Chinese defense contractors also have good prospects in Malaysia, Algeria and several other markets. At present, about 50 countries from all over the world are involved in arms trade or defense industry cooperation with China.

Pakistan

Pakistan is China's largest defense customer. It buys more Chinese weapons than any other country and also partners with China on various weapons programs. The nature of China-Pakistan relations in this area is quite similar to the relations between Russia and India.

Cooperation between Beijing and Islamabad also has the hallmarks of a strategic alliance directed against India. That alliance came into being after the war between India and China in 1962, and became firmly established by the early 1970s. In the 1980s (during the Soviet campaign in Afghanistan), the Sino-Pakistani alliance was also aimed against the Soviet Union. Yet another factor that stimulated arms trade between the two countries is Islamabad's tense relations with the West, especially on the issue of the Pakistani nuclear program. Suffice it to recall that the country was under a US arms embargo from 1990 to 2003.

Pakistan now prefers to pursue joint weapons development programs with China, rather than just buying finished Chinese products. The largest of these joint programs are:

- Joint development and manufacturing of the FC-1 fighter (designated in Pakistan as the JF-17 Thunder). Pakistan provided 50% of the R&D financing for the program. In 2006-2008, it took delivery of eight aircraft assembled in Chengdu. In March 2009 China signed an agreement on joint production of the FC-1 (JF-17) for the Pakistan Air Force with the state-owned Pakistan Aeronautical Complex (PAC). The first JF-17 assembled in Pakistan was rolled out at PAC in Kamra on November 23, 2009, and some 26 units had been assembled by the autumn of 2011. In May 2011, China's AVIC signed a new contract, worth an estimated $1 billion, to supply another 50 JF-17 assembly kits to Pakistan. By 2020, the Pakistan Air Force plans to take delivery of 250 units. Some of the JF-17s may be exported as well. The bulk of the components for the aircraft assembled in Pakistan are Chinese, but the RD-93 engine is supplied by Russia via China.

- Joint production of the K-8 trainer. Pakistan contributed 25% of the R&D financing for this program. From 1994 to 2003, Islamabad took delivery of 12 aircraft, then from 2006 to 2011 assembled 27 K-8P units from Chinese components at PAC under a 2005 agreement with China. There have been reports that an additional batch of K-8Ps may be made for the Pakistan Air Force.

- Joint development of the Al Khalid main battle tank and manufacturing in Pakistan. The Al Khalid is based on the export version of the Chinese Type 90-II (MBT-2000) main battle tank, which is a modified copy of the Soviet T-72. China and Pakistan signed the contract for the development of the tank in 1990. Mass production began at Heavy Industries Taxila in Pakistan in 2000. Most of the components, including the 125mm tank gun, are Chinese. The engine and transmission are supplied by Ukraine's Malyshev Plant. China delivered 10 assembly kits in 2006 and another 18 in 2007. A total of 320 tanks have been ordered by the Pakistan Army, of which about 200 have already been produced. There have been reports that a contract for another 250 tanks could be signed. Pakistan is also trying to find foreign buyers for the new tanks. Before the Al Khalid, Pakistan used to assemble Type 85-IIAP tanks using Chinese components.

- Production of F-22P frigates in Pakistan under Chinese license. The first three Project F-22P ships were built for Pakistan in China under a 2005 contract. The fourth is now under way at Karachi Shipyard & Engineering Works, for delivery in 2013. The shipyard itself is being modernized with Chinese assistance. It is possible that more of these ships will be built in Pakistan under Chinese license. Karachi Shipyard & Engineering Works is also building a Chinese-designed 500 t missile boat. The first such boat was launched in China in 2011.

In addition, the Institute of Industrial Control Systems in Pakistan makes Anza series man-portable SAM systems under Chinese license. The Anza Mk I, which entered production in 1989, was a licensed copy of the Chinese HQ-5 (which was itself a copy of the Soviet 9K32 Strela 2). The Anza Mk II (a copy of the Chinese QW-1) was launched in 1994, and the Anza Mk III (a version of the QW-2 with some local improvements) in 2008. Other weaponry made in Pakistan under Chinese license includes the HJ-8 anti-tank missile system (designated as the Baktar Shikan in Pakistan), the Type 83 122mm MLR system (Azar, a clone of the Soviet BM-21 Grad), small arms, anti-tank weapons, ammunition, etc.

China is also providing assistance to Pakistan in the development of missile technologies. Several Pakistani ballistic missiles are believed to have been developed with help from the Chinese, including the Hatf 3 (Ghaznavi, a version of the Chinese M-11), which has a range of up to 300 km.

Pakistan also buys some of the most advanced weapons systems designed and manufactured in China itself; indeed, for several of these weapons systems it was the launch customer. The biggest recent contracts include:

- A 2009 contract with the AVIC aerospace corporation for 36 J-10 fighters made in Chengdu, worth $1.4 billion, for delivery in 2014-2015. The Pakistani designation of the J-10 is the FC-20. This is China's first export contract for the J-10.
- A 2008 contract with AVIC for four ZDK-03 AWACS aircraft, worth $278 million, for delivery in 2010-2014. The first aircraft was delivered in 2011.
- A 2005 contract for four Project F-22P frigates, worth $750 million. Three of them were built by Hudong Zhonghua Shipbuilding in Shanghai and delivered in 2009-2010. The fourth is now on the way in Pakistan. China also supplied six Z-9C naval helicopters for these frigates.
- A 2008 contract of 36 units of the A100 300mm MLR system (a clone of the Soviet Smerch MLR), for delivery in 2011-2012. Pakistan also ordered the SLC-2 artillery radar systems as part of the contract.
- A 2006 contract for 300 PL-12/SD-10 medium-range air-to-air missiles for Pakistan's FC-1 (JF-17) fighters. These new missiles were designed in China and are now manufactured there with Russian help.

In recent years, Pakistan has also purchased at least 110 S-802 anti-ship missiles and at least 11 YLC-2 and YLC-6 air defense radars. In 2003-2004, it bought 143 Type 96 122mm towed howitzers (a clone of the Soviet D-30). There have been reports that it now plans to buy SH-1 155mm self-propelled howitzers made by China's CNGC. Finally, China has offered to supply diesel-electric submarines; the two countries are now negotiating a contract for up to six boats.

On the whole, Chinese arms deliveries to Pakistan, including deliveries under the FC-1 program, could be worth as much as $500 million to $600 million a year. This constitutes 33% to 50% of Chinese arms exports for the period until 2015.

Finally, Pakistan's role is not limited to being the biggest buyer of Chinese weapons. For China, the country also serves as an important conduit for Western defense technologies.

Iran

Chinese arms exports to Iran are much less transparent than sales to Pakistan. Over the past two decades, Tehran has been trying to achieve complete self-sufficiency in terms of weapons, so its cooperation with China is limited mostly to the acquisition of defense technologies, licensing programs or joint R&D. Deliveries of finished Chinese weapons systems to Iran have been few and far between since 2000. We have no information about any deliveries of combat aircraft, tanks or other big weapons platforms.

There is, however, some evidence of extensive Sino-Iranian defense industry cooperation in areas such as the manufacturing of several missile systems in Iran under Chinese license. The list of Chinese missiles made in the Islamic Republic includes:

- FL-8 anti-ship missiles (designated as Kosar in Iran), as well as the FL-9 (Nasr), C-801 (Tondar), C-802 (Noor) and C-803 (Ghader);
- FM-80 SAM system (Shahab Thaqeb, also known as the Ya-zahra, a copy of the French Crotale);
- HQ-2 (Sayyad-1, a modified copy of the Soviet S-75 SAM system);
- QW-1 (Misagh-1) and QW-2 (Misagh-2) man-portable SAM systems.

Iran also makes Chinese artillery systems and small arms, as well as the Boragh tracked APC (based on the Chinese Type 86, which is a clone of the Soviet BMP-1, but without a gun turret).

From 2001 to 2004, China built seven China Cat missile boats for Iran and transferred a license for them. Tehran has also purchased Chinese JL-14 air defense radars. Finally, Iran appears to have been the launch customer for China's latest CM-802AKG air-to-surface missile; deliveries commenced in late 2010.

On the whole, however, the scale of Sino-Iranian defense industry cooperation should not be exaggerated. We estimate that over the period of 2002-2009, Iran received a mere $200 million worth of Chinese weaponry. But Iran is becoming increasingly isolated from the West, and lately from Russia as well, following Moscow's cancellation of the Iranian contract for the S-300PMU2 SAM systems. Beijing therefore remains Tehran's only source for more or less modern weaponry and defense technology. It would be entirely logical for Iran to seek closer defense industry cooperation with China, including imports of advanced weapons platforms (especially aircraft). It is not clear, however, to what extent China will be ready to accept Iranian overtures in the face of Western pressure.

Burma

Burma (Myanmar) became one of the biggest buyers of Chinese weapons after the military regime came to power in 1989, which triggered Western sanctions. In the early 1990s, Burma imported large quantities of Chinese weaponry, although most of it was obsolete, and some was received as a gift of aid. Nevertheless, these deliveries did much to improve the overall level of weapons technology used by the Burmese army, which had hitherto been one of the world's most poorly equipped. Since 2000, Burmese arms imports from China have fallen (the country has started to buy from other suppliers as well, including Russia), but the weapons it buys are more advanced.

China remains one of the main suppliers of weapons to Burma. In the West, Burma has long been seen as a de facto satellite of Beijing, but now the Burmese military regime is clearly trying to pursue a balanced foreign policy; this applies to its arms import strategy as well.

Like many other isolated regimes, Burma regards the development of its national defense industry as a high priority. It pursues active cooperation with China in this area, and regards Chinese technologies as the best option—politically, economically and technologically. The two countries are now working on joint aerospace and shipbuilding projects.

In 1999, Burma bought 12 Chinese K-8 trainers, and in 2009 it signed a contract with AVIC for another 60. Twelve units were delivered in 2010; another 48 are to be assembled from Chinese components in Burma itself, where China is now helping build the production facility. The deal includes technology transfer.

In 1998, China built the hulls of three Anawrahta (Sinmalaik) class corvettes for Burma. They were then completed and fitted out with modern Western systems by the Burmese at their own Navy shipyard in Sinmalaik. The same shipyard has been building, with Chinese assistance, two Chinese-designed frigates. Also with help from China, the Burmese are building missile and patrol boats, for which China is supplying weapons and various systems. Some 60 Chinese C-801 anti-ship missiles were supplied for Burmese fast attack missile boats and corvettes in 2004-2005.

In 2009-2010, it was reported that Burma had purchased a large batch (at least 50) of the Chinese VT1A (MBT-2000) tanks. Other Chinese imports include army trucks, communications systems, radars, etc.

Egypt

Egypt established military ties with China after the Sadat regime finally broke with the Soviet Union in the late 1970s. Since then Cairo has been buying advanced weapons systems mainly from Western countries. But it did recognize China as an alternative source of large batches of cheap and simple weaponry, as well as a supplier of parts and components to maintain the Soviet hardware purchased in previous years. As a result, Egyptian arms imports from China rose quite significantly from 1976 to 1985.

However, Cairo gradually lost interest in obsolete and unreliable Chinese weapons. In the past two decades, arms trade between the two has fallen sharply, and now appears to be limited to Chinese supplies of parts and components, transfer of some technologies and a few joint defense industry projects.

The biggest Egyptian contract for Chinese weapons was signed in 1999, when Cairo agreed to pay China's HAIC (now part of the AVIC group) $500 million to launch assembly of K-8E trainers in Egypt. Cairo took delivery of 10 finished planes; another 110 were assembled in Egypt in the years 2001-2009 from kits supplied by HAIC. The two countries are now negotiating a similar arrangement for the FC-1 fighters.

Sudan

China has been selling arms to Sudan since the late 1960s; in the 1980s, it became one of the leading military partners of the Sudanese government. Sudan, which has come under Western sanctions and pressure, views China as a strategic partner and a source of affordable weaponry, which the government badly needs to fight the unending ethnic conflicts on its territory. The country is reported to have bought $83 million worth of Chinese weapons in 2005. We estimate that its aggregate arms imports from China from 2001 to 2010 total more than $1 billion.

Sudan buys mostly small arms and infantry weapons, ammunition and trucks from China. But there have also been some contracts for heavy weaponry and aircraft. There is very little information available, but in the past decade Sudan is reported to have received 20 Chinese A-5C fighter-bombers (in 2003); 12 K-8 trainers (2005-2008); at least 10 Type 85-IIM tanks; 50 FN-6 man-portable SAM systems (in 2006); and 10 WZ-551 APCs (in 2002-2004, but deliveries in this category were probably made in other years as well). In recent years, China seems to have ceded its position as Sudan's biggest supplier of heavy weapons and aircraft to Russia and the CIS republics.

North Korea

China has been North Korea's key military ally ever since the arrival of Communist governments in the two countries. China and the Soviet Union have been the two biggest arms suppliers to Pyongyang. China's assistance has played an important role in North Korea's efforts to develop a powerful and self-sufficient defense industry. After the breakup of the Soviet Union, China became Pyongyang's sole ally and supporter in every single area, including defense.

Because of North Korea's extreme secrecy, there is no information about the current state of military-technical cooperation between the two countries. But that cooperation is known to be ongoing; this was confirmed by officials in October 2010. As far as we can tell, in the past two decades, defense industry ties between China and North Korea have involved mostly the transfer of technologies and some joint R&D projects. Imports of finished weapons systems are probably quite limited, as North Korea does not have the money to pay for them; in any event, the country is pursuing a policy of seeking complete self-sufficiency in arms procurement.

3.4 Other Defense Customers and Partners

Algeria

Over the past decade, Algeria has been actively buying Chinese weaponry, although some of the contracts are not being disclosed. The country is known to be working with China on building warships. Beijing helped build three Djebel Chenoua class corvettes. Sometime after 2000, it supplied various components and weapons for those ships, including 25 C-802 anti-ship missiles. China also built the *Soummam*, a large training ship for the Algerian Navy, under a 2004 contract. On the whole, however, imports from China play a minor role in Algeria's arms procurement programs. The country prefers to buy more advanced systems from Russia or Western Europe. But the situation may well change as Chinese defense technology continues to improve. There have been reports that Algeria has shown interest in the FC-1 fighters.

Angola

In 2007 China and Angola signed an agreement on military-technical cooperation. It has been reported that as part of the agreement, Angola has been receiving Chinese small arms, military equipment and army trucks. The country is now regarded as one of China's key economic and political partners in Africa. Given that Angola is relatively wealthy, it could potentially become one of the biggest buyers of Chinese weaponry on the continent. There have been reports that Angola is interested in the FC-1 fighter, and that in 2010 it was negotiating a possible contract for K-8 trainers.

Argentina

In recent years, Argentina has faced major economic problems, and money is still tight. It has therefore shown increased interest in Chinese weapons. It has discussed the possibility of assembling the Z-11 light helicopters on its territory, and an agreement to that effect was signed in October 2011. It has also discussed a possible contract for Chinese radars. In 2010, it took delivery of four WMZ551B1 wheeled APCs as a pilot batch, with a view to buying more if the vehicles perform well.

Armenia

Armenia purchased four Chinese WM-80 237mm long-range MLR systems under a 1998 contract. Additionally it may have bought small arms and infantry weapons.

Azerbaijan

There have been reports that since 2008, Azerbaijan has been showing interest in buying 24 FC-1 fighters assembled in Pakistan.

Bangladesh

Bangladesh has been a traditional Chinese defense customer since the mid-1970s, when most of the equipment in service with the country's armed forces was of Chinese origin. For Bangladesh, cooperation with China has political motivations as well as military ones: the country wants to distance itself from India, and it maintains close relations with Pakistan. But Bangladesh is a poor country, and the main attraction of Chinese weapons for Dhaka is the price. Some of the weapons it imports come from Chinese Army surplus, or are given by China as a gift of aid.

From 2007 to 2010, Bangladesh took delivery of 12 Chinese F-7BG fighters and five FT-7BG aircraft, plus modern rocket weaponry for them (PL-7 and PL-9C missiles) under a 2002 contract worth $118 million. Since 2000 it has bought 54 Type 96 122mm towed howitzers (upgraded clones of the Soviet D-30); 20 Type 83 122mm howitzers; HJ-8 anti-tank missiles; NH-5JA1 and QW-2 man-portable SAM systems; and large numbers of mortars, grenade launchers, small arms, ammunition and trucks.

Bangladesh purchased Chinese weapons and components for fitting out the frigate *Bangabandhu*, which was built in South Korea, and for upgrading a Jianghu-1 class frigate purchased from China itself. These weapons included C-802 missiles and HQ-7 SAM systems. Bangladesh is thought to be one of the most likely future customers for the FC-1 fighter. Finally, in 2010, it was negotiating a contract for two Project F-22P frigates and three Z-9C naval helicopters. In 2011, Bangladesh also ordered 16 of the new F-7BG fighters.

Bolivia

Bolivia has always been a cash-strapped country, and its army is rather archaic. It the early 1990s, it started to buy cheap Chinese weapons—mostly for financial reasons. It bought 18 Type 66 152mm towed howitzers (a copy of the Soviet D-20) from Chinese Army surplus, as well as 36 Type 54-1 122mm howitzers (a copy of the M-30) and 28 NH-5A man-portable SAM systems. In 2001, it also purchased 50 HJ-8 anti-tank missile systems and a batch of Chinese RPGs. Cooperation with China was stepped up following the election in 2005 of President Evo Morales, a left-wing radical known for his anti-Western rhetoric. In 2008, Bolivia used a Chinese loan to buy two MA60 transport aircraft. In 2009, it bought six K-8 trainers worth $57.8 million after securing favorable payment terms, with a number of installments spread over a long period. In June 2011, it signed a contract with AVIC for another two MA60 transports for the Bolivian Air Force and six H425 helicopters (a commercial version of the Z-9); the deal appears to have been financed by a Chinese loan, too. Finally, the country has purchased several batches of Chinese trucks.

Cambodia

China was the main sponsor of the Khmer Rouge during their rule in Cambodia in 1975-1978, as well as during their later insurgency. China now supports the current government in Phnom Penh and has established close military ties with it. Cambodia receives Chinese weapons, equipment and trucks under various aid programs. Since 2005, it has taken delivery of 16 Type 062-1 patrol boats, one assault-landing boat and a 60 m floating dock from Chinese Navy surplus. Several patrol boats were purpose-built for Cambodia in China. In August 2011 Cambodia signed a $195 million contract for Z-9 helicopters. That has probably been the Chinese helicopter industry's biggest export contract to date.

Chad

Chad recently launched an ambitious rearmament program. After establishing diplomatic relations with China in 2006, it immediately proceeded to place orders for Chinese weaponry. In 2007, it took delivery of 10 ZFB05 armored vehicles. It likely buys Chinese small arms, infantry weapons and ammunition as well. It can be expected that Chinese sales to Chad will grow.

Democratic Republic of the Congo

Under former president Mobutu Sese Seko, who ruled Zaire from 1965 to 1997, the country imported large amounts of weaponry from China. In recent years, Beijing has been providing limited military assistance to the Democratic Republic of the Congo. In 2007, it began to supply trucks, weapons and military equipment. In 2010, the DRC was reported to have expressed interest in the new L-15 trainer. In November 2010, it signed a contract for six MA60 passenger transports and two Y-12 light transports worth $150 million, paid for by a Chinese loan.

Ecuador

The country has pursued closer military ties with China since the arrival of the left-wing Correa administration in 2005. China has supplied two BT-6 (CJ-6) piston-engine trainers, trucks and other military hardware as a gift of aid. Ecuador has bought HJ-8 anti-tank missile systems and paid for more trucks. In 2008, it signed a $60 million contract for two YLC-2V and two YLC-18 radars for its air force. Finally, in 2009 it ordered four MA60 passenger transports worth $52 million; the money came from a $438 million Chinese loan.

Ghana

Ghana has bought significant numbers of Chinese weapons in recent years. From 2007 to 2009, it signed contracts for six K-8 trainers, two MA60 passenger transports, and two Y-12 light transports. In 2005, it bought 55 army trucks; and in 2009, 48 wheeled ZFB05 APCs. In 2009, it placed an order for four large patrol boats, which were supplied in 2011.

Indonesia

From 1965 to 1997, when the Suharto regime was in power, Indonesia pursued distinctly anti-Chinese policies and clearly saw China as its main external threat. In the post-Suharto period, Indonesia has gradually become involved in various cooperation projects with China in many areas, including arms trade. The main attraction of Chinese weapons for Indonesia is price, plus flexible financing options. In addition, at times Indonesia has had to face Western and American pressure as well as sanctions, forcing it to look for non-Western sources of arms supplies.

In 2005, Indonesia bought large numbers of the QW-1 and QW-3 manportable SAM systems. In 2008, it signed a contract for two batteries of the TD-2000B self-propelled SAM systems equipped with the new QW-4 missiles. The Indonesian Navy has bought at least 23 C-802 anti-ship missiles and an undisclosed number of C-705 light anti-ship missiles. There are plans for more C-802 contracts. For the new missile boats now under way in Indonesia, the country plans to buy Chinese clones of the AK-630 30mm shipborne artillery systems. On the whole, Chinese arms exports to Indonesia will likely continue to grow, but the Chinese are unlikely to break into any new segments of the Indonesian defense market.

Jordan

Being a relatively poor country with scarce natural resources, Jordan has in recent years turned to China as a source of affordable infantry weapons. From 2005 to 2007, it bought 150 W86 120mm mortars, 375 WW90 60mm mortars and 1,275 PPT89 RPGs. Chinese arms exports to the country will likely continue to grow.

Kenya

Although the country maintains close economic and political ties with the West, over the course of the past 15 years it has increasingly been turning to China for weapons supplies. It bought 12 Y-12-II light transports from 1997 to 2000; 32 WZ551 wheeled APCs in 2007; and eight Z-9 helicopters in 2010, four of which were the Z-9WA attack version. Kenya is now considering the possibility of buying the K-8 or L-15 combat trainers.

Kuwait

Between 1998 and 2001, Kuwait signed two widely publicized contracts for Chinese weapons worth a total of $387 million. Under those contracts, from 2001 to 2003 it received 51 of the latest PLZ45 155mm self-propelled howitzers, 51 PCZ45 transport-loaders built using the same chassis, and 10 armored command-and-control vehicles. The contract was clearly political, an attempt by Kuwait to cozy up to a permanent UN Security Council member. The country's contract for the Russian BMP-3 infantry fighting vehicles and the Smerch MLR systems likely pursued similar goals. So far, there have not been any further Kuwaiti contracts for Chinese weaponry.

Malaysia

Malaysia was hostile toward China and Chinese influence for a long time, but that attitude seems to have changed since the turn of the century. The country has always tried to diversify the sources of its arms imports, so it is not ruling out cooperation with China in that area. Chinese suppliers have submitted bids for several Malaysian defense contracts. For now, however, the only actual sale they have closed is a batch of 60 FN-6 man-portable SAM systems delivered under a 2008 contract. In addition, Malaysia is looking into the possibility of buying longer-range Chinese SAM systems such as the KS-1A.

Mexico

Arms trade between China and Mexico began only recently, and so far is negligible. In 2006, Mexico bought 13 M-90 105mm towed mountain howitzers (a copy of the Italian M56). It was reported that the Mexicans were unhappy with the quality and decided not to order any more howitzers (the initial plan was to buy 96).

Morocco

Although Morocco is not a traditional Chinese customer, in recent years financial woes have caused the country to turn to China for affordable weapons. It has reportedly signed three large contracts, but there has been no official confirmation. It is said that Morocco became the launch customer for China's latest VT1A (MBT-2000 tanks) after signing a contract for 150 units in 2008. Deliveries commenced in 2009. There have also been reports about Moroccan contracts for 20 K-8 trainers and 18 A100 (Smerch clone) 300mm MLR systems. One final detail worth noting is that Russia must have been in the running for the Moroccan tank contract with its T-90S, but lost to the Chinese. It can be assumed that arms trade between China and Morocco will continue.

Namibia

China has been one of Namibia's leading arms suppliers ever since the African country gained independence. In 1997, Namibia received two Y-12 light transports. In 2001, it took delivery of four K-8 combat trainers; it is now discussing a contract for another eight units. From 2006 to 2008, it received 12 F-7NM and two FT-7NM fighters under a contract signed in 2005. In 2009, China delivered 21 wheeled APCs, which are now in service with the Namibian UN force. In 2010, it was reported that Namibia was showing interest in the new L-15 trainer.

Nigeria

Nigeria has been buying Chinese weapons since the 1970s. In 2005, it signed a $251 million contract for 12 F-7NI and three FT-7NI fighters, plus air weapons (including 20 PL-9C missiles). Deliveries were completed in 2010. Nigeria also buys Chinese small arms and ammunition.

Oman

Oman has been buying weapons from both Russia and China since 2000, in an apparent effort to diversify its arms import sources. In 2002-2003, it bought 50 WZ551 wheeled APCs from China and a battalion (18 units) of the Type 90 122mm MLR systems (a version of the Soviet BM-21 Grad). There have been no reports of any further contracts since then.

Peru

Chinese arms exports to Peru have been ongoing since the early 1990s; in recent years sales have picked up. China is close to becoming the leading arms supplier to the country. Since 2005, Peru has bought HJ-8 anti-tank missile systems and trucks. In 2008-2009, it placed an order for 25 FN-6 and five QW-18 man-portable SAM systems. Also in 2009, it announced

China the winner of a contract for 120 tanks. The Chinese were to supply their VT1A model (MBT-2000) equipped with Ukrainian 6TD-2 engines. Five of these tanks were delivered, but the rest of the contract fell through due to financial reasons and Ukraine's refusal to supply the engine and transmission blocks.

The Philippines

The Philippines has demonstrated growing interest in military cooperation with China, despite the ongoing political tensions between the two countries (including territorial disputes in the South China Sea). One of the main reasons for that interest is that Manila cannot afford to spend too much on arms procurement. Starting in 2007, Beijing has been providing military assistance worth several million dollars a year (mainly in the form of non-combat and engineering equipment). Manila has also discussed the possibility of buying eight Z-9 helicopters.

Rwanda

In recent years, China has become one of the leading defense partners of Rwanda. It trains Rwandan specialists and supplies various weaponry. It delivered six Type 86 or 96 (Soviet D-30) 122mm towed howitzers in 2007 and 20 ZFB05 armored vehicles in 2008. There have also been reports that the country is discussing a contract with China for the F-7MG fighter.

Saudi Arabia

The first Saudi contract for Chinese weapons was signed in 1986, triggering something of a stir on the international scene. The kingdom bought 12 DF-3A medium-range (up to 2,800 km) missile systems with a set of 60 ballistic missiles. In an effort to diversify the sources of its weapons imports, Saudi Arabia has been showing increased interest in non-Western weaponry since 2000. In 2007, it bought 54 PLZ45 155mm self-propelled howitzers (identical to the ones bought by Kuwait). Deliveries were made in 2008-2009. There have also been reports that the country has purchased a batch of 155mm towed howitzers from China. Both countries are secretive about their arms trade, so it cannot be ruled out that other contracts between them have been signed as well.

Sri Lanka

Sri Lanka has been buying Chinese weapons since the early 1970s. During the long Tamil Tigers insurgency from 1981 to 2009, China was one of the main suppliers of weapons to the government. Sri Lanka purchased large batches of Chinese small arms, artillery and ammunition, as well as aircraft. Since 2000, China has supplied four F-7GS fighter jets (delivered in 2008), nine K-8 trainers, 10 BT-6 (CJ-6) piston-engine trainers, and two Y-12-IV light transports. In 2010, Sri Lanka ordered six MA60 transports and another two K-8 trainers. It has also bought radars, WZ551 wheeled APCs, and some other equipment. Following the rebels' military defeat in 2009, it appeared likely that Sri Lankan arms imports would fall sharply and that imports from China would be affected as well. Nevertheless, in October 2010 it was reported that Sri Lanka was planning to buy advanced fighter jets, with the Russian MiG-29 and China's FC-1 (possibly assembled in Pakistan) being viewed as the prime candidates.

Syria

Syria began to pursue military ties with China after the breakup of the Soviet Union. But Syria's buying power is limited; in addition, for a long time China was unable to offer any weaponry advanced enough to interest Damascus. Arms trade between the two therefore remains negligible, although Beijing is thought to have supplied the country with several military technologies, including missile technologies. US sources estimate that Syria has bought about $1 billion worth of Chinese weapons since 2000.

It has been reported that in the late 1990s, Syria imported an unknown number of M-9 (DF-15) tactical missiles with a range of up to 600 km, as well as 20 M-11 (DF-11) missiles with a range of up to 350 km. It later launched production of its own version of the M-11, with a 280 km range. The rest of the contracts with China were likely for parts, components and technology transfers. It can be assumed that as Chinese weaponry becomes more advanced, Syria will become more interested, and arms trade between the two countries will pick up.

Tanzania

Tanzania is one of China's traditional African defense customers. The first shipments of Chinese weaponry were made in 1969, and most of the weapons in service with the Tanzanian armed forces are Chinese. In the past two decades, however, arms trade between the two countries has declined. In 1994, Tanzania bought two Y-12-II light transports, and two Y-8-F200 transports in 2004. In 2006-2007, China supplied four WZ551 wheeled APCs. In 2007, the two countries signed an agreement under which China undertook to provide assistance in equipping the Tanzanian Air Force and in training Tanzanian pilots. It appears that the two F-7 fighters China delivered in 2009 (probably from its Air Force surplus) were supplied under the terms of the 2007 agreement. There have also been reports that the two countries are negotiating a contract for K-8 combat trainers.

Thailand

Thailand became one of China's biggest defense customers in the early 1980s. Shared hostility toward Vietnam forged closer military-political ties between the two countries. Large shipments of simple and affordable Chinese weaponry continued from 1985 to 1992. In later years, however, arms trade between the two countries declined, owing to changes in the political situation in Southeast Asia and the world. Besides this, Thailand itself had developed interest in systems more advanced than anything the Chinese could offer at the time.

Nevertheless, China remains one of the leading players on the Thai defense market, and has retained a strong presence in some segments. It built two Pattani class patrol ships for the Thai Navy under a 2002 contract worth 80 million euros; more contracts for these ships could be in the pipeline. In 2000 Thailand bought fairly large batches of C-801 and C-802 anti-ship missiles, and it is now considering the possibility of buying coastal defense systems equipped with the C-802. Finally, China offered to supply one or two Ming class diesel submarines from its surplus, but Thailand eventually declined, viewing the boats as too obsolete.

Turkey

Ankara has pursued military cooperation with China since the early 1990s, mostly in the form of licensing Chinese technologies and participating in joint R&D projects. China has provided assistance in the development of Turkish surface-to-surface missiles. Under the terms of a deal signed in 1997, Turkey purchased 19 batteries of the WS-1 tactical rocket systems (with a 100 km range) and licensed their upgraded version, the WS-1B, which has been made since 2006 by Turkey's Roketsan under the local designation T-300 Kasirga. China has also helped Turkey launch mass production (mostly at Roketsan facilities) of the more advanced J-600T Yildirim tactical missile system with a range of up to 150 km. The design is based on the Chinese B-611 missile. Several sources report that since 2001, Turkey has bought between 15 and 18 B-611 missiles as a finished product. The first demonstration of the Yildirim took place during a military parade in Ankara in 2007. Cooperation of this kind will likely continue. It has been reported that in recent years the Chinese have invited Ankara to take part in several joint R&D programs in areas such as air defense, armored vehicles and UAVs.

There have been no reports of any direct Turkish imports of finished Chinese weapons systems, apart from the aforementioned missiles. Chinese defense contractors are trying hard to break into the Turkish market. For example, they bid for a Turkish contract for medium- and long-range SAM systems with their HQ-9.

Uganda

Uganda buys Chinese small arms and equipment. In 2002, it purchased 102 army trucks. In 2008-2009, it took delivery of four Y-12-IV light transports. In recent years, Uganda has sped up its rearmament programs, so it may show growing interest in affordable Chinese weaponry.

Ukraine

Ukraine supplied the AI-222K-25 engines, made by Motor Sich, for China's L-15 combat trainer program. There are now plans to launch production of the L-15 in Ukraine itself; the aircraft can be used by the Ukrainian Air Force as a light fighter and attack aircraft. Negotiations have been ongoing since 2007. In September 2010, President Viktor Yanukovich spoke in favor of a project to launch joint production of the L-15 in Ukraine, but the feasibility of that project is unclear.

United Arab Emirates

The United Arab Emirates has been buying small quantities of Chinese weaponry since the 1980s. In 1993-1994, it received 20 Type 59-1 130mm towed guns (a copy of the Soviet M-46). During the same period, it also bought several HY-4 coastal defense anti-ship missile systems, despite their obvious obsolescence.

Venezuela

Chinese defense cooperation with Venezuela began long before the arrival of President Hugo Chávez, with his staunch anti-American policies. The president himself seems to regard arms trade with China as a counterbalance to arms trade with Russia. It is therefore safe to expect that Venezuela will increase its arms imports from China.

The first big Venezuelan contract for Chinese weapons was the purchase of three JYL-1D air defense radars in 2005. The country later placed an order for another 12 JYL-1 series radars worth $150 million; deliveries commenced in 2010. Under a contract signed in 2008, Venezuela purchased 18 K-8 trainers, and in 2010, it announced that it had placed an order for another 22. In 2011, Venezuela signed a contract for eight Y-8 medium transports (a clone of the An-12).

Zambia

Zambia has traditionally been one of China's key defense customers in Africa. Most of the weapons in service with the Zambian armed forces are Chinese; from time to time, the country receives small batches of Chinese weaponry as a gift of aid. Most of the contracts signed over the past 15 years have been for Chinese aircraft. In 1996-1997, Zambia bought five Y-12-II light transports. In 2006-2007, it placed an order for five Y-12-IV aircraft, plus two MA60 passenger transports (paid for by a Chinese loan). In 2000, the country took delivery of eight K-8 trainers; there were reports that Zambia intended to buy another eight. Zambia is also thought to be a potential customer for the Z-9 helicopter and the FC-1 fighter. In 2010 it was reported that the country was showing interest in the latest Chinese fighter jet, the J-10.

Zimbabwe

Zimbabwe has been a large recipient of Chinese weapons since the arrival of the black majority government in 1980. The Mugabe regime, which is now facing strong international pressure and Western sanctions, is increasingly dependent on China as its main supplier of weaponry. In 2005-2006, Zimbabwe took delivery of 12 Chinese K-8 trainers. In 2008, it received a batch of mortars, anti-tank weapons, small arms and ammunition. The country is thought to be a potential customer for the FC-1 fighter and the new L-15 combat trainer.

3.5 Chinese Weapons Available for Export

China has become one of the world's leading weapons manufacturers. The Chinese defense industry offers almost the entire range of weaponry used by the world's armed forces, although many of its products are technologically dated, rely on imported components or suffer from problems with quality. This is why Chinese arms exports are competitive only in certain segments of the global defense market. Customers who want advanced high-tech platforms still have to turn to suppliers from the developed countries. Nevertheless, China's recent progress in defense technology and industrial capability is obvious, so the competitiveness of Chinese weapons on the global market will continue to increase.

At present, the Chinese defense industry offers the following weapons categories for exports.

Combat aircraft

The foundations of the Chinese aerospace industry were laid during the 1950s and 1960s, thanks to massive Soviet assistance. During that period, China launched production under Soviet license (or simply copied) such aircraft as the Il-28 (Chinese designation H-5) and Tu-16 (H-6) bombers; MiG-17 (J-5), MiG-19 (J-6) and MiG-21F-13 (J-7) fighters; An-2 (Y-5), Il-14 (Y-6), An-12 (Y-8) and An-24 (Y-7) transports; Yak-18 (CJ-5/6) trainers; and Mi-4 (Z-5) helicopters. Until fairly recently, those aircraft made up the bulk of the Chinese aerospace industry's defense output. Indeed, upgraded versions of some of those planes, such as the J-7, Y-7, Y-8 and CJ-6, are still being made.

In later years, amid growing isolation from both the Soviet Union and the West, China failed to make much progress, resorting instead to endless iterations of the venerable MiG-19 (J-6) and MiG-21 (J-7) designs. Both types of aircraft were exported; an upgraded version of the J-7 is being made to this day under export contracts. In fact, it is the world's last second-generation fighter still in production. China's attempts at developing a third-generation fighter (J-8) and a heavy fighter-bomber (JH-7) proved largely disappointing. Real progress began only in the late 1980s, first with Western assistance, then with Russian help after 1991. Since then, China has made a real breakthrough in combat aircraft, with the results of that breakthrough starting to appear during the past decade.

In recent years, China has managed to bring to mass production status three important programs started in the early 1990s: the J-10 fourth-generation multirole fighter; the FC-1 fourth-generation light fighter, developed primarily for export; and the JH-7 attack aircraft. Ongoing R&D projects include the J-20, a fifth-generation fighter, the first prototype of which took to the air in early 2011.

J-7 fighter

Initially, the J-7 was the Chinese version of the MiG-21F-13 made under Soviet license. Production began in 1967. For a long time the J-7 remained China's most advanced combat aircraft. The Chinese have continued to make their own improvements over the past 40 years, and developed several new modifications. Up to 3,000 units of the J-7 have been manufactured in China to date. Between the early 1970s and the mid-1990s, up to 500 units of the F-7 (export designation) were supplied to Albania, Bangladesh, Burma, Egypt, Iraq, Iran, North Korea, Pakistan, Sri Lanka, Sudan, Tanzania and Zimbabwe.

A new wave of export contracts for the J-7 family of fighters came in the past decade, after China developed an updated version with an improved wing and better radars, electronics and weapons. In 2003, the PLA took its first deliveries of the J-7G version, which remained in production simultaneously with the latest J-10 fighter until 2008. About a hundred of the J-7G and J-7GB fighters were delivered to the Chinese Air Force. Production of the JJ-7 two-seat trainer for the PLA continued until the autumn of 2010. From 2001 to 2003, Pakistan took delivery of 57 F-7PG fighters fitted with Italian Grifo MG radars, plus nine FT-7PG two-seaters. Between 2006 and 2008, deliveries were made to three foreign customers: Bangladesh (12 F-7BG and four FT-7BG), Namibia (12 F-7NG and two FT-7NG), and Sri Lanka (five F-7SG and one FT-7SG). In 2010, Nigeria received 12 F-7NI and three FT-7NI aircraft under a 2005 contract.

Finally, the JJ-7 was used as a basis for the development of the JL-9 (FTC-2000), a supersonic trainer jet that entered a flight test program in 2003 and is now being offered to the Chinese Air Force.

J-8 fighter

The twin-engine J-8 was China's first attempt at developing an domestic fighter. The program was launched in the early 1960s. It made use of the Mikoyan Design Bureau's documentation for the E-152, an experimental Soviet interceptor. Small-series production of the J-8 began only in 1979. The aircraft was essentially a supersized twin-engine version of the J-7; its design was obsolete, and in general it proved a disappointment. A total of about 100 units (in several modifications, including a reconnaissance version) had been made by the time production ceased in 1987.

In the 1980s, the aircraft was completely redesigned and turned into something similar to the old Soviet Su-15 interceptor. The new J-8II series entered small-series production in 1995. There have been several versions of it (J-8B, J-8D, J-8H and J-8F), each representing an incremental improvement on the last. The J-8F entered production in 2003. The Chinese developed a reconnaissance version, the JZ-8F, and worked on an attack version, the J-8G, geared to take on the enemy's air defenses. It is believed that a total of up to 300 J-8II series aircraft have been manufactured for the Chinese Air Force, and small-series production will likely continue for a few more years. In 1996 the Chinese unveiled the J-8IIM, an export version of the fighter equipped with Russian electronics, including the Zhuk radar. So far no export contracts have been signed for that aircraft, although since 2004 the Chinese have advertised versions of it fully outfitted with domestically made equipment. In 2009, they unveiled a new modification, the J-8T (F-8T), equipped with the same avionics as the latest J-10 fighter. The aircraft is being marketed primarily to foreign customers.

J-10 fighter

The domestic J-10, a light single-engine multirole fighter, has been China's most important national aerospace program in the past two decades. In the coming years the J-10 should become one of the main types of combat aircraft serving with the Chinese Air Force, along with the Russian Su-27 (J-11) and Su-30 fighters. It is believed that the J-10 program was kick-started between 1988 and 1991, when China signed an agreement with Israel Aircraft Industries (IAI) to acquire the documentation for the Lavi fighter program, which had been discontinued in Israel itself under US pressure. The J-10 is believed to have inherited Lavi's aerodynamic design. Since 1994, Russian companies have also been heavily involved in the J-10 project. All the J-10s in mass production today are equipped with the Russian AL-31FN engine (maximum thrust of 127.5 kN). The Chinese have also considered using the Russian Zhuk-10PD radar for their new fighter jet. In addition, Russia is providing assistance in the development of guided airborne weapons for the J-10.

The J-10A version entered mass production in 2003. A total of 250 units, including prototypes, are thought to have been built by the autumn of 2011. The J-10AS two-seat combat trainer entered production in 2006. In an effort to fully localize the aircraft, the Chinese have developed a domestic turbofan engine, the WS-10A Taihan, which will be used on the J-10B modification. But the supplier is still struggling with teething problems with the new engine, so China remains dependent on imports of the AL-31F family of engines from Russia. The AL-31FN's thrust is clearly insufficient for the J-10, which is too heavy for a single-engine aircraft, so the Chinese will likely try to secure the supplies of Russia's latest Type 117S engine.

The J-10 is already being offered to foreign customers. Pakistan has signed a contract for 36 aircraft (designated as the FC-20) to be delivered in 2014-2015. Iran and Venezuela are also reported to have shown interest. But China's ability to export the J-10 is limited by the need to secure supplies of the AL-31FN engine from Russia, which insists on vetting any re-exports.

J-11B, J-15 and J-16 fighters

The J-11B is a pirated copy of the Russian Su-27SK fighter (Chinese designation J-11A), which China assembled under Russian license from 1998 to 2004. The J-11B is believed to share much of its electronics and its radar with the J-10. It is equipped with two Chinese WS-10A engines. Three J-11B prototypes underwent flight tests in 2006; production for the Chinese Air Force was launched in 2008. According to various sources, 50 to 100 units had been built by late 2011; it is possible that some of them were equipped with imported AL-31F engines. The Chinese have also developed a two-seat version of the aircraft, the J-11BS, as well as the J-16 two-seat multirole fighter, a direct copy of the Russian Su-30MK2. In the future China intends to offer the J-11B to foreign customers, but for now any large contracts seem unlikely due to China's continued dependence on imported Russian components.

In 2009, China built the first prototype of the J-15 carrier-based fighter, which is very similar to the Russian Su-33. The J-11B design was used as a starting point. The aircraft is equipped with the WS-10A engine and is now undergoing flight tests. It appears that the designers of the J-15 have borrowed some of the solutions used in the Russian T-10K-7 fighter, a prototype of the Su-33. Beijing bought one such aircraft from Ukraine in 2003.

FC-1 fighter

The light multirole FC-1 (Fighter China-1) Xiaolong fighter is a product of China's ongoing J-7 (MiG-21) modernization program. Development of the FC-1 began in cooperation with America's Grumman (now part of Northrop Grumman) as part of the Super-7 program. But Sino-American defense industry cooperation ended after the Tiananmen Square protests. In 1991, the project to develop a new light fighter jet (destined primarily for exports) by fundamentally upgrading the J-7 was resumed by China on its own, with the new FC-1 designation. Pakistan became one of the main potential customers after the country came under US sanctions in 1990 and could no longer buy the American F-16s. As with the J-10, Russian specialists have been involved in the FC-1 program since the early 1990s. The designers of the FC-1 have reportedly used many solutions developed by the Mikoyan Design Bureau for its Type 33 light tactical fighter, which was in development prior to 1986. China had decided right from the start to use many imported (primarily Russian) components, such as the engine and radar. In 1998, Pakistan joined the FC-1 program, contributing about half of the R&D costs.

Flight tests of the first FC-1 prototypes began in 2003. The aircraft entered mass production in 2006 under Pakistani contracts (Pakistani designation JF-17 Thunder). From 2006 to 2008, Islamabad took delivery of eight aircraft. In 2009 and in 2011, China signed agreements with the Pakistan Aeronautical Complex on joint production of 92 FC-1 (JF-17) fighters at PAC facilities for the Pakistani Air Force. The first Pakistani-made JF-17 was rolled out at a PAC factory in Kamra on November 23, 2009. By the autumn of 2011, PAC had assembled 26 units; the production rate had reached two aircraft per month. Pakistan plans to take delivery of up to 250 fighters for its own air force by 2020, as well as to win foreign customers for the aircraft. It appears that most of the systems and components for the JF-17s assembled in Pakistan are supplied from China. The Chinese Air Force clearly has no plans to buy any FC-1 fighters, and likely views the program as backup in case of any problems with the J-10.

As with the J-10, the FC-1 program relies heavily on imports of Russian engines. All the FC-1s now being made are equipped with the Russian RD-93 turbofan engine, a modification of the RD-33. Chinese re-exports of the RD-93 to Pakistan are a delicate matter for Moscow, given Russia's close ties with India. The JF-17s built under the Pakistani Air Force contracts are to be fitted out with Chinese radars. The export versions of the aircraft will use other radars, including the Kopye-F developed by Fazotron-NIIR. China is now working to equip the FC-1 with an domestic turbofan engine, the WS-13 (which appears to be a copy of the RD-93). The first flight of an FC-1 prototype fitted with the WS-13 took place in March 2010.

The FC-1 is regarded as having the potential to be a hit on the world market for combat aircraft. It is the cheapest fourth-generation fighter currently on offer, and relatively affordable for poorer countries. It can certainly give the used MiG-29s offered by the former Soviet republics a run for their money. The price of one FC-1 is estimated at approximately $20 million, or even less. Among those named by Chinese officials as potential customers are Iran, Egypt, Morocco, Lebanon, Bangladesh, Sri Lanka, Zimbabwe, Zambia, Malaysia, Nigeria and Azerbaijan. The success of the J-10 and FC-1 programs has therefore laid a solid foundation for China's transformation into a big exporter of relatively advanced combat aircraft.

JH-7 fighter-bomber

The JH-7 heavy attack aircraft has been in the making for a very long time, serving as a symbol of China's inability to build world-class combat planes.

The twin-engine, two-seat fighter-bomber is China's answer to the Su-24 and the Tornado. The program was launched in 1973, but was held back by problems with the development of airborne radars and electronics, as well as the absence of a suitable Chinese engine. The testing of the first prototype began only in 1988. By 1993, a total of four prototypes and 20 mass-produced units had been built. After further improvements the Chinese made another 20 or so modified JH-7 aircraft from 2002 to 2004, but production was then discontinued because the design had become too obsolete. The aircraft were equipped with Rolls-Royce Spey Mk 202 turbofan engines with a maximum thrust of 93 kN; a total of 140 engines were purchased.

Since the mid-1990s, the Chinese have made some radical changes to the JH-7 design. They have replaced almost all the electronics with modern systems made in China with Russian and Western assistance; many electronic components are shared with the J-10. The aircraft's fighting ability has been augmented by guided weapons. The new modification is designated as the JH-7A. It is equipped with WS-9 engines (a copy of Spey Mk 202 made in China under British license), although the thrust specifications are clearly insufficient for such a heavy aircraft. The JH-7A entered mass production in 2005; a total of 180 units were made by the end of 2011. Production will likely continue for some years to come.

China is now marketing two export versions designated as the FBC-1 Flying Leopard (JH-7) and FBC-1M Flying Leopard II (JH-7A), but has not secured any contracts so far. The aircraft is not a very attractive proposition, so it is unlikely to win any customers.

H-6K bomber

China continues to manufacture the H-6 long-range bomber (a copy of the old Soviet Tu-16). Recently it launched production of the H-6K modification equipped with Russian D-30KP2 turbofan engines. Exports of this aircraft are unlikely.

Combat trainers

Chinese Air Force cadets begin their training on the CJ-6, a piston-engine aircraft and a modification of the CJ-5 (a licensed copy of the Soviet Yak-18A), which China used to make in the 1950s. The CJ-6 series entered mass production in 1962; a total of 1,800 units have been produced so far, some of them destined for export under the PT-6 designation. The plane has long since reached obsolescence, but China continues to produce it in small batches.

As a replacement for the CJ-6, China launched the joint L-7 program (Russian designation Yak-152K) with the Russian Yakovlev Design Bureau in 2006, using the Yak-152 design as a starting point. The first prototype, equipped with the Russian M-14Kh piston engine, was completed in 2010. Several sources have reported that the Chinese Air Force plans to buy up to 300 L-7 trainers and offer the aircraft to foreign customers. It cannot be ruled out that some of the Chinese-made L-7s will be purchased by Russia itself.

China's main combat trainer to which cadets graduate after the CJ-6 is the JL-8 (K-8) jet. The first deliveries of the JL-8 to the Chinese Air Force were made in 1998. An estimated 200 units have been manufactured so far, all equipped with AI-25TLK turbofan engines made by Ukraine's Motor-Sich. The Chinese have also developed the JL-11 version equipped with the domestically made WS-11 (a copy of the Soviet-designed and Slovak-made DV-2F), but it does not appear to have entered mass production. The export version of the JL-8 is designated as the K-8.

The advanced trainer aircraft to which trainees graduate before they become proper pilots is the MiG-21U two-seater, which was used in a similar capacity by the Soviet Air Force. The aircraft was in production in China from 1987 to 2010, designated locally as the JJ-7. Well aware that the aircraft had long reached obsolescence, in the early 2000s China announced a competition for a replacement. Two designs were offered: the JL-9 and the L-15. The JL-9 (JJ-9, FTC-2000) is an incremental upgrade of the JJ-7. It took off on its maiden flight in 2003; a total of 10 prototypes and pre-production units were made. The L-15 (JL-15) was designed with the assistance of Yakovlev Design Bureau, using the Yak-130 as a starting point. The first prototype took to the air in 2006; three more had been built by the end of 2010. The L-15 is equipped with AI-222K-25 engines supplied by Ukraine's Motor-Sich. There are also plans for the L-15AJT version (equipped with the same engines) to be used for basic and advanced pilot training, and for the L-15LIFT, a supersonic version which can also be used as a light combat aircraft. The latter will be equipped with the AI-222K-25F afterburning engine. One of the four prototypes made so far is the L-15LIFT version. Finally, there are plans to develop a light single-pilot fighter based on the L-15LIFT design.

The Chinese Air Force must soon make a choice between using the JL-9 or the L-15 as their primary trainer jet. The L-15 is being energetically marketed to foreign customers. Pakistan, Egypt, Venezuela, Namibia, Zimbabwe and the DRC have all expressed interest. China has also been in talks with Ukraine about the possibility of local production of the L-15.

Transports and special-purpose aircraft

China no longer makes the Y-7 transport, which was a copy of the Soviet An-24. But its successor, the MA60, which entered production in 2000, is a direct descendant of the Y-7 design. The new passenger transport can hold 50-60 people on board, and was developed for both civilian and military applications. The MA60 is equipped with Pratt & Whitney Canada PW127J turboprop engines and Western avionics. China is actively marketing the aircraft to foreign customers. Out of the more than 160 units for which orders have been placed so far, 60 are destined for export. On the whole, however, reception of the new aircraft has been mixed.

The modified MA600 version entered production in 2009. The Chinese are now developing the MA700 regional turboprop airliner, which has 70- and 80-seat configurations. The MA700 appears to be a clone of the European ATR 72 (China makes fuselage sections for ATR aircraft as a subcontractor).

The Y-8, a four-engine turboprop transport, is a copy of the Soviet An-12B. It entered production in 1980, and in 1991 China launched the upgraded Y-8F version with new avionics. More than a hundred Y-8 transports had been manufactured by 2010. Venezuela placed an order for eight in 2011. The "Westernized" Y-8F-600 version, equipped with Pratt & Whitney Canada PW150B engines and Western avionics, has been in development since 2000, but its future is uncertain. The Y-9 transport, a descendant of the Y-8 now being developed with the help of Ukraine's ANTK Antonov, can lift 20 t of payload. The first prototype took to the air in 2011.

In the light transport category, China continues to make the Y-12, a domestically designed twin-engine turboprop aircraft that entered production back in 1982. The Y-12 is fitted with Pratt & Whitney Canada PT-6A engines. Thanks mainly to its affordability, it has been well received by foreign customers; out of the 150 units manufactured so far, 110 have been sold abroad. The first prototype of the new Y-12F version was built in 2010; up to 30 units have been ordered so far.

AVIC is now working on the Y-20 heavy transport, which has been in development since 1993. In 2007, the program was designated as an industry priority by the Chinese State Council. The starting point for the Y-20 design was the Russian Il-76, which has been in service with the Chinese Air Force for decades. Ukraine's ANTK Antonov is also involved in the program. The designers hope that the Y-20's specifications will be similar to those of the American C-17 Globemaster II. Premier Wen Jiabao spoke of the importance of this program during a tour of AVIC facilities in 2008. Officials said the maiden flight of the Y-20 would take place in 2012, but it cannot be ruled out that the deadline will be pushed back. The first mass-produced units will be equipped with Russian D-30KP2 engines, which will later be replaced with the Chinese WS-18 turbofan (the engine is still in development). The WS-18 is a version of the CFM56, a joint product of General Electric and Snecma. As part of the Y-20 program, AVIC plans to complete an ambitious retooling program at its Xi'an subsidiary (Xi'an Aircraft Industry Company).[100]

The Chinese aerospace industry has yet to launch its first heavy aircraft platform, so it is still facing problems with the development of special-purpose aircraft. It has used the obsolete Y-8 design to build eight patrol aircraft and a certain number of reconnaissance, electronic warfare and command-and-control aircraft. China has three different models of aircraft equipped with airborne warning and control systems (AWACS): the KJ-2000 (based on the Il-76); the KJ-200 (Y-8); and the ZDK-03, a version designed specifically for export to Pakistan and apparently also based on the Y-8.

Helicopters

The first helicopter manufactured in China was the Z-5, a licensed version of the Soviet piston-engine Mi-4. Production began in 1959, and a total of 545 units had been made by the 1980s. In the mid-1960s, the Chinese tried to develop their first domestic helicopter, the Z-6. It was essentially an enlarged and modified version of the Z-5 with a single WZ-5 turboshaft engine (a variant of the WJ-5, which was a copy of the Soviet AI-24A). Superficially, the Z-6 looked similar to the first single-engine prototypes of the Soviet Mi-8. But the Chinese never managed to get it to work properly. The Z-6 program was cancelled in 1979 after only 15 units were built. There had been other programs in the 1960s and 1970s, including the medium-sized Z-7 and two light models, the Yan'an and Model 701. Some of those designs were left on paper, some reached the stage of prototypes, but none entered mass production.

As a result of these failures, China was unable to launch the next generation of helicopters with gas turbine engines until it bought the SA321 Super Frelon transport from France's Aerospatiale and acquired a license in the 1980s for the AS365N2 Dauphin 2 medium helicopter. With Aerospatiale's help, the Chinese launched production of the Dauphin 2 (Chinese designation Z-9) and later the Super Frelon (Z-8). They also copied the Aerospatiale AS350B Ecureuil light helicopter (Z-11).

Although China's political relations with the West remain strained, the country's helicopter industry maintains close cooperation with European companies. As part of a deal with Eurocopter, the Chinese launched production of the Eurocopter EC120 light helicopter (Chinese designation HC120) in 2005. They have also begun joint development of the EC175 (Z-15) medium helicopter. Eurocopter provided assistance in designing the frame of the Z-10 attack helicopter, and Italy's AgustaWestland aided in developing the gearbox. China has signed agreements with European partners to launch production of the AgustaWestland AW109 (Chinese designation CA109) and Schweizer 300 light helicopters. It has also signed a deal with Poland to assemble W-3 and SW-4 helicopters. The turboshaft engines for Chinese helicopters are sourced from Europe, the US and Canada. On the whole, however, the Chinese helicopter industry is still in its early stages. One measure of its competence will be whether or not it can successfully bring the Z-10 and Z-15 programs to mass production.

Beijing has always been an eager buyer of the Russian Mi-8/Mi-17 series helicopters; in fact, most of the units exported in the past 10 years were destined for China. In 2007, China began assembling Mi-171 helicopters from components supplied by Russia's Ulan-Ude aircraft plant. The first unit was assembled in Sichuan in December 2007, but since then the program seems to have stalled. Reports claim that the biggest customer for the locally assembled helicopters is the PLA. The Chinese are now discussing with Russian companies the possibility of jointly developing a new heavy transport helicopter (presumably based on the Russian Mi-46 design).

Z-8 multirole helicopter

The Z-8 is a versatile machine with a maximum take-off weight of 13 t. It is a copy of the Aerospatiale SA321 Super Frelon, a three-engine transport; China bought 13 units from France in the 1970s. Starting in 1976, the

Chinese first tried simply to pirate the French design, but later on signed a cooperation agreement with Aerospatiale. Small-series production of the Z-8 began in 1989; up to 60 units were built by the end of 2010. Twenty units had been delivered to the Chinese Navy by 1999, fitted out as transports, anti-submarine, and search and rescue helicopters. All of them were equipped with WZ-6 engines, a copy of Turbomeca's Turmo IIIC made under French license in China. About twenty Z-8A transport and assault-landing helicopters have been made for Chinese Army Aviation since 2002; they are equipped with imported Turbomeca Makila 2A engines. A similarly equipped search and rescue version for the Chinese Air Force (designated as the Z-8K) entered production in 2007; at least 20 units have been made. Since 2004, the manufacturer has also built at least six Z-8JH and Z-8S search and rescue helicopters for the Chinese Navy; they are quite similar to the Z-8K. Finally, in 2004 it completed the first prototype of the civilian Z-8F version equipped with Pratt & Whitney Canada PT6B-67A engines. Since then, the Z-8F design has been used to develop the AC-313 civilian helicopter; the first unit took to the air in 2010.

Z-9 multirole helicopter

The versatile Z-9 helicopter is a licensed Chinese version of the Aerospatiale (now Eurocopter) AS365N2 Dauphin 2. Its maximum take-off weight is 4 t. The Z-9 is equipped with two WZ-8 engines, which are a copy of Turbomeca's Arriel engine made under French license in China. Production of the Z-9 in China from French assembly kits began in 1981; a total of 28 units were made. China then made 20 partially localized Z-9A units. The fully localized Z-9A-100 version entered production in 1993. The Chinese are now working on the new Z-9B version; its main rotor is made of composite materials. Almost all of the Z-9 units made so far have been ordered by the PLA.

The Chinese have developed several specialized military versions of the Z-9, including the WZ-9G and Z-9W attack versions; the Z-9WZ reconnaissance version; the Z-9C anti-submarine version and the Z-9D attack version, both carrier-based; the DZ-9 electronic warfare version; the Z-9AC2 aerial spotting helicopter for use with artillery units; the Z-9S and Z-9KA search and rescue helicopter; an airborne signal relay version; and a trainer. Small-series production of the H410A and H425 versions for civilian Chinese customers began in 2005. The Chinese are now marketing another civilian version, the AC312. A total of 300 Z-9 series helicopters were made in China by the end of 2010. In recent years, Beijing has secured several export contracts for these machines.

Z-10 attack helicopter

The Z-10, a twin-engine, tandem-cockpit attack helicopter also designated as WZ-10, is China's first domestically designed and developed helicopter—either military or civilian. The Z-10 development project also involved Eurocopter and AgustaWestland. The first prototype took to the air in 2003. The Chinese initially built at least six units equipped with imported Pratt & Whitney Canada PT6C 67C engines. In 2010, they launched small-series production of a new version equipped with domestic WZ-9 turboshaft engines. There have been reports that the Chinese are using the Z-10 frame, engine and gearbox to develop the new CMH medium multirole helicopter.

Z-11 multirole helicopter

The Z-11 has a maximum take-off weight of 2.2 t. It appears to be a pirated copy of the Aerospatiale (Eurocopter) AS350B Ecureuil model, although the Chinese insist that the Z-11 is a completely domestic design. The helicopter entered mass production in late 1997. It is equipped with WZ-8 engines made under French license in China. An estimated 70 units have been built for military and civilian customers, including 43 Z-11J trainers delivered to the PLA. The Chinese have also developed the Z-11WA reconnaissance and attack version. Ignoring protests from Eurocopter, China is marketing the Z-11, including the civilian AC311 modification, to foreign customers. In 2009, it announced a project to develop the AC102 light civilian helicopter based on the Z-11 design and equipped with Honeywell LTS101-700D-2 engines.

Z-15 multirole helicopter

The Z-15, which has a maximum take-off weight of 7 t, is being developed mostly for civilian applications. This is a joint program with Eurocopter (European designation EC175) that was launched in 2005. The helicopter is equipped with two Pratt & Whitney Canada PT6C-67E engines. It can also be fitted with two Turbomeca Ardiden 3C engines, which China plans to produce locally under French license (Chinese designation WZ-16). The first prototype took to the air in France in December 2009. The first deliveries of European-made units are to commence in 2012; Chinese-made units (destined primarily for China's own market) will arrive in 2014. By early 2010, orders for 114 units had been received, including 30 from Russia's UTair airline.

Unmanned aerial vehicles

China is currently developing a broad range of unmanned aerial vehicles (UAVs) of all classes and types. More than 70 different Chinese companies displayed UAV-related exhibits at an aerospace industry event in Xi'an in June 2009. It is not clear, however, how close these projects are to mass production, and whether any of the products displayed in Xi'an are now in service with the PLA.

In 1981, the Chinese Air Force took delivery of its first WZ-5 UAV, a long-range airborne reconnaissance system equipped with a jet engine and weighing 1,700 kg. The WZ-5, which China continues to improve upon, is a copy of the old American AQM-34N Firebee system. It was developed by the Beijing University of Aeronautics and Astronautics (BUAA), which has since become one of the leading Chinese designers of UAVs. The university is now part of the AVIC aerospace group, as are all the other research institutes mentioned in this section below.

China's first tactical reconnaissance UAV was the fairly primitive D-4 (ASN-104/105), a 170 kg craft equipped with a piston engine. It was developed by the Northwestern Polytechnical University (NWPU) and went into production in 1985. In 2009, China demonstrated the BZK-006 (WZ-6), a 250 kg tactical UAV that was also developed by NPU and has already entered mass production. It is similar to several Israeli-made craft of this type; the design is based on the ASN-206 project, which was launched in the 1990s. China is also advertising the 320 kg ASN-209 and the 480 kg ASN-207, both developed by NPU. Another modern UAV in service with the PLA is the man-portable and hand-launched ASN-15, which weighs only 6 kg. Other NPU designs that rely on piston engine technology include the Ba-2, Ba-7 (ASN-7) and Ba-9 (ASN-9).

In 2005, the PLA took first deliveries of the W-30 (18 kg), W-50 (95 kg) and PW-1 (130 kg) tactical UAVs, as well as the Z-1 (9 kg) rotorcraft. All of them were designed by the Nanjing Research Institute of Simulation

Technique (NRIST). Chinese UAVs offered to foreign customers include the PW-2 (220 kg) and two unmanned mini-helicopters, the Z-2 (35 kg) and Z-3 (100 kg). In 2008, China's Research Institute No. 611 demonstrated the CH-3, a 640 kg attack UAV which can carry pod-mounted weapons. In 2006, another Chinese R&D institute that specializes in police technology presented a similar attack UAV, the Blue Eagle. China's CATIC corporation is marketing the U8E, a 220 kg unmanned helicopter; the SH-1 and SH-3 tactical UAVs; and the TF-8 and TF-10 hand-launched mini-UAVs. The Beijing Institute of Aerodynamics has developed three LT-series micro-UAVs; all three have likely entered production at this time. Another manufacturer, the Beijing Wisewell Avionics Science and Technology Company, is offering the AW-series mini-UAV, which weighs less than 10 kg.

BUAA and GAIC have jointly developed the DCK-007 (Sunshine), a 750 kg reconnaissance UAV which can climb to medium altitudes and stay airborne for long periods of time. Several units have already been delivered to Chinese customers. Another joint BUAA-GAIC project, the BZK-005, is more technologically advanced and weighs 1,250 kg. Several units are now in service with the Chinese Air Force. Other advanced BUAA designs include the Huifeng tactical UAV and the DUF-2 micro-UAV.

In 2003, China began testing the BZK-009 (WZ-9), a high-altitude, long-range UAV powered by a jet engine and weighing 7,500 kg. This GAIC design is similar to America's Global Hawk. Previously GAIC had also advertised the WZ-2000 project, which is a heavy UAV equipped with a jet engine and airborne weapons. The project may still be ongoing under the Luoyang designation. GAIC and Institute No. 611 have jointly developed the Soaring Eagle, a high-altitude UAV that is now being tested. The CAC company in Chengdu has developed the Xianglong, a 7,500 kg UAV of the Global Hawk class; the craft entered a test program in 2009. In 2008, CAC also began testing the Yilong, a medium-altitude 1,150 kg UAV equipped with a piston engine. Another CAC design is the Tianyi, an 80 kg tactical UAV.

Finally, China is reportedly developing several long-range reconnaissance and attack UAVs similar to America's Predator. In this category in 2010, AVIC demonstrated the Pterodactyl 1, while XAC and NPU showed the ASN-229A. Chinese designers are also working on the Warrior Eagle project, which is a low radar profile attack UAV powered by a jet engine. In 2010,

CASC corporation demonstrated the WJ-600, a medium reconnaissance and attack UAV which has reportedly entered mass production, and the SL-200, a high-altitude reconnaissance craft.

Airborne and anti-ship missiles

For a long time China lagged far behind the world leaders in airborne weapons technology, but it has made rapid progress in this area since the turn of the century. Much of that progress has been based on imports from Russia and direct Russian technological assistance. Despite its recent achievements, Beijing continues to buy large amounts of airborne weaponry from Russia and Ukraine.

In the air-to-air missiles sector, China for a long time had no alternative but to copy old Soviet short-range missile technology. Starting in the mid-1960s, its main short-range missile was the PL-2, equipped with an infrared targeting system. The missile was a copy of the Soviet R-13, which was itself a copy of the American AIM-9B. In the mid-1980s, China launched production of the PL-5, a modified version of the old PL-2. For a long time the PL-5 family remained China's primary short-range missile; it was also very popular among China's defense customers abroad. In 2008, China unveiled the latest modification of that missile, the PL-5E-II. In 1987, it had also launched production of the PL-7, which is a copy of the French Magic R.550. In the early 1990s, it started making the PL-8, a version of the Rafael Python 3 made under Israeli license.

In the mid-1990s, China launched production of its most advanced short-range missile yet, the PL-9, which is based on an improved PL-5 design and employs the all-aspect seeker technology used in the PL-8. China continues to improve the PL-9 (the new PL-9C version entered production in 1999) and also uses it in its SAM systems. The PL-9C and the new short-range missile China is now developing, the PL-10 (first tested in 2008), appear to use a seeker developed by the Arsenal Design Bureau in Kiev. Chinese attack helicopters are armed with the TY-90 (PL-90) short-range air-to-air missile originally developed for SAM systems.

All Chinese air-to-air missile development and manufacturing assets have been consolidated within the AVIC aerospace corporation. China's first medium-range air-to-air missile was the PL-11 (FD-60), a licensed copy of the Italian Aspide missile equipped with a semi-active radar seeker. Before the launch of the PL-11, the Chinese had unsuccessfully tried to copy the Aspide's American prototype, the AIM-7. They acquired a license for the Aspide in 1986, but managed to bring the PL-11 (as well as the SAM systems that use that missile) into mass production only in 2000. China's new-generation PL-12 missile (SD-10) is equipped with an active radar seeker. All the main components of this missile were developed by Russian subcontractors: the seeker by the Agat Research Institute in Moscow, the control system by Vympel, and the inertial targeting system by the Tikhomirov Instrument Design Institute. The PL-12 is believed to have entered mass production in 2007; it will be exported to Pakistan as part of a contract for FC-1 fighters. It appears that the homing devices for the missile are imported from Russia. There have also been reports that China is developing a longer-range version of the PL-12 with a ramjet engine. It is likely that Russia is involved in that project as well.

China did not have any tactical air-to-surface missiles in its arsenal until it secured deliveries from Russia in the late 1990s. Sometime after 2000, China and Russia's Zvezda-Strela launched joint production on Chinese territory of the Russian Kh-31P anti-radar missile, designated in China as the YJ-91 (KP-1). All the components of this missile, including the homing heads, were imported from Russia. After 2005, China launched production of an domestic air-to-surface missile, the KD-88. The missile is equipped with a turbojet engine and has a range of over 100 km. It uses a TV guidance system, and there may also be an anti-radar version. In the past two years China has announced the completion of several other airborne missile projects. These include the compact AR-1, which weighs only 45 kg; the BA-7 (47 kg), and the TV-1; all three are equipped with a semi-active laser seeker and can be used by helicopters and UAVs.

China has developed, and continues to improve upon, a broad range of air-launched anti-ship missiles based on similar shipborne designs. These include the C-701 (YJ-7), C-704, C-705, FL-6, FL-8, FL-9, TL-2 (JJ-2), TL-6 (JJ-6), and TL-10 (JJ-10) light missiles, all of them designed primarily for export. In the medium class, China offers the C-801K (YJ-8K), C-802K (YJ-82K) and C-803K (YJ-83K) missiles. It is now developing the supersonic YJ-12 anti-ship missile. China's H-6 long-range bombers are armed with C-601 (YJ-6) and C-611 (YJ-61) anti-ship missiles, and the KD-63 (YJ-63) air-to-surface cruise missile. The first two are descendants of the Soviet P-15; they are equipped with a turbojet engine. The KD-63 cruise missile is based on the C-611 design. China's strategic aviation is armed with the CJ-10A. The missile entered mass production a few years ago and is similar to the Soviet Kh-55; the Chinese designers have apparently used the Kh-55 design as a template.

The development of the C-801 series of anti-ship missiles (which use solid-fuel technology) and of the C-802/C-803 missiles (turbojet engines) has been an important achievement for the Chinese defense industry. The ship-borne versions of the C-801 and C-802 have been quite popular with many third world buyers over the past 15 years. In 2010, China demonstrated the CM-802AKG air-to-surface missile, which is based on the C-802 design. It is equipped with an infrared seeker with terminal guidance capability.

In recent years, China has also been actively improving its guided airborne bombs. Here, too, it seems to rely on Russian technologies. The LT-2 500 kg laser-guided bomb (LS-500J, GB1, similar to the Russian KAB-500L) entered service in 2006. China is now testing an upgraded version, the LT-3, which is equipped with a satellite guidance system. It has also announced the completion of several other satellite-guided bomb projects, including the FT-1 (500 kg), FT-2 (a version of the FT-1 with wings), FT-3 (250 kg) and FT-5 (100 kg). Finally, it has developed the LS-6 gliding and satellite guidance module for 50, 100, 250 and 500 kg bombs. The 250 and 500 kg versions can also be equipped with an additional laser guidance system.

Ships

The unprecedented rise of China's shipbuilding industry over the past decade has yet to translate into a commensurate growth in Chinese defense exports in this segment. China's share of the world market for warships has long been negligible, and after 1990 it actually began to shrink even further. The main reason is that China has yet to close the large gap between itself and world leaders in such crucial technology areas as naval weapons systems, radars, sonars, combat control systems and naval propulsion units. Between the 1960s and the 1980s, the Chinese could still compete with their simple and affordable attack boats (including those copied from Soviet designs) in the developing countries' markets. But by the 1990s, the Chinese shipbuilding industry had become uncompetitive even in that segment. From 1994 to 1996, China built 10 Houdong class missile boats (a version of the old

Soviet Project 205) for Iran and another three for Yemen—but they were a very dated piece of technology. In the past decade, the Chinese have managed to sell only a handful of such boats to foreign customers. They are, however, advertising more up-to-date products, such as the China Cat, a 23 m missile boat; several sources report that seven such boats were built for Iran from 2001 to 2004. In 2002, China delivered five patrol boats to Iraq under a contract signed back when Saddam Hussein was still in power. In recent years, Chinese shipyards have been building patrol boats for several third world countries, such as Cambodia, East Timor and Ghana.

Contracts for larger and more complex Chinese warships have been few and far between, coming mostly from poor countries or from China's political allies. In the mid-1970s, China delivered seven Type 033 medium diesel-electric submarines to North Korea. Another 20 boats were later built in North Korea itself with Chinese assistance. (The Chinese Type 033 is a copy of the old Soviet Project 633, which entered service in the 1950s.) Another four Type 033 submarines were delivered to Egypt in the 1980s. Chinese Type 053H frigates have been exported to Thailand (four units), Egypt (two), and Bangladesh (one, from the Chinese Navy surplus). By the 1980s, Type 053H had become hopelessly obsolete, so in 1994 Thailand placed an order for two Type 25T frigates (designed primarily for foreign customers). Most of the weaponry and electronic systems for the two ships were imported by Thailand from the US and installed at its own shipyards.

On the whole, Chinese exports of surface ships were very modest in the 1990s and 2000s. In 1998, China built the hulls of three 1,200 t Anawrahta (Sinmalaik) class corvettes for Burma. The Burmese then had these hulls fitted out with modern Western systems and weapons in Burma itself. In 2005-2006, China built two 1,400 GT Pattani class patrol ships for Thailand.

The Chinese shipbuilding industry's biggest foreign contract so far has come from Pakistan, which ordered four Project F-22P frigates (an export version of Type 053H3) worth 750 million dollars in 2005. All the weapons and electronics on these ships are Chinese. The first three ships were delivered in 2009-2010. The fourth is now under way at Karachi Shipyard and Engineering Works, with a delivery date set for 2013. The Karachi shipyard itself is being upgraded with Chinese assistance, and more frigates may be built in Pakistan under Chinese license in the future. In 2010, Islamabad also signed a contract for two 500 t missile boats of a new type. One of them was completed by China's Tianjin Xingang Shipbuilding Heavy Industry Company in 2011; the other is under way in Karachi.

In 2010, the Chinese were discussing a possible contract for two Project F-22P frigates with Bangladesh.

China has also secured several foreign contracts for auxiliary ships. The largest order, for the *Similan* fleet replenishment supply ship, came from Thailand. The ship was delivered in 1996. China also built the *Soummam*, a large training ship, under a 2004 contract with Algeria.

Almost all the ships supplied to foreign countries from the Chinese Navy surplus have been given to China's allies in Africa and Asia as gifts of aid. In recent years the largest recipient was Cambodia, which since 2005 has received 16 Type 062 patrol boats.

China recently launched several new types of non-nuclear submarines, but has not sold any to foreign buyers as yet. That may soon change; for example, a possible contract for six submarines is now being discussed with Pakistan.

In summary, China has seen its exports grow in certain segments, such as naval weapons (primarily C-801 and C-802 anti-ship missiles). But China has yet to close the gap with world leaders in many areas of naval technology, and for this reason exports of Chinese warships remain very modest. It is not clear whether China can change that situation over the coming decade. Competition in this market is fierce; to make matters worse for the Chinese shipbuilders, many Western navies have decommissioned large numbers of older ships and are now offering them to foreign buyers. It is therefore quite likely that the bulk of the China's shipbuilding contracts will continue to come from China's own navy, with export contracts playing a very minor role.

It is safe to assume that over the coming decade, exports of Chinese warships and boats will be limited to traditional Chinese customers such as Pakistan, Bangladesh and Thailand, as well as poor countries in Asia and Africa, for whom price is the most important factor.

Air defense systems

China is developing a very broad range of air defense systems, but its most technologically complex programs in this area are still facing serious challenges. The country is lagging well behind the world leaders in air defense technology, and continues to import advanced systems from Russia. Russian companies are also involved in most of China's ongoing R&D programs in this area.

Apart from the 28 battalions of S-300PMU-1 and S-300PMU-2 systems bought from Russia, China's main air defense system is the HQ-2, which is a copy of the old Soviet S-75 (SA-2). Up to 50 HQ-2 battalions are now in service with the PLA. During the period between 1966 and the 1990s, China developed many HQ-2 modifications; the most advanced are the HQ-2J and the HQ-2P. An domestic descendant of the HQ-2 is the HQ-12 (KS-1), armed with solid-fuel missiles with a range of up to 50 km. The KS-1A version remains in production to this day, but no more than 10 battalions have actually been deployed since the late 1990s.

China's most advanced medium- and long-range SAM system is the HQ-9, which went into production in the early 2000s. The design is similar to the Soviet S-300PMU, and was developed with Russian assistance. The specifications are also close to the S-300PMU, and the range is over 120 km. The Chinese seem to be facing difficulties with mass production of the new system; no more than 10 battalions had been deployed as of the autumn of 2010. Nevertheless, the HQ-9 has been offered to Turkey and Pakistan. China has also developed a simplified version of the system, designated as the FT-2000. Its missile uses a passive radar seeker and its primary intended target is AWACS aircraft. A similar version of the missile, the FT-2000A, has been developed for use with the HQ-2 SAM systems.

China's latest medium-range SAM system, the HQ-16, was developed with Russian assistance and uses the Russian 9M317-series (SA-17) vertical-launch missiles. It is not clear whether the system has entered mass production. In 2008, China demonstrated a prototype of the LS-II, a mobile SAM system armed with Chinese PL-12 (SD-10) medium-range air-to-air missiles. The PL-12 has a range of up to 30 km and is equipped with an active radar seeker, which was developed with Russian assistance. The LS-II SAM system can also use PL-9C short-range (up to 15 km) air-to-air missiles equipped with an infrared seeker. The Chinese have also developed a version of the system which uses containerized vertical-launch PL-12 (SD-10) missiles.

The first Chinese short- and medium-range SAM systems were the domestically developed HQ-61A and the HQ-7 (FM-80 and FM-90); both have a range of up to 12 km. The HQ-7 was a copy of the old French Crotale system. The FM-90 modification, with a range of up to 15 km, is still in production. It is China's main short-range SAM system, and is also being offered to foreign customers. In the late 1990s, a batch of FM-90s was sold to Iran, which went on to copy the design. In the 1990s, the Chinese developed the HQ-64 (LY-60) short-range SAM system. It uses the Aspide air-to-air missile made under Italian license in China, and has a range of up to 18 km. A limited number of these systems have been delivered to the PLA. The Chinese and Pakistani armed forces have also received small numbers of towed and self-propelled (mounted on the WZ551 APC chassis) DK-9 SAM systems armed with PL-9C short-range air-to-air class missiles (15 km range) with an infrared seeker. Since 2000, the Chinese have also been advertising the TY-90 (Yitian) SAM system; its missiles have a range of up to 6 km and are equipped with an infrared seeker. These systems are offered in towed and self-propelled versions (the latter uses APC and truck chassis), but so far only a small number of them have been delivered to the PLA (all of which were the towed SG-II version), and none have been sold abroad. The bottom line is that in the sector of advanced short-range SAM systems, China still relies on Tor-M1 systems bought from Russia; domestically developed Chinese systems are still being produced only in small numbers. Several joint projects with Russia are known to be under way in this area.

China's first man-portable SAM system was a copy of the Soviet 9K32 Strela-2 (SA-7), which went into production in China in the 1970s under the local designation NH-5. Its various versions remained in production until quite recently. It was succeeded by the NH-6 (FN-6). A relatively small number of these systems have been made since the 1990s; some have been exported. In 2008, China demonstrated an improved version of the system, the FN-16. The next generation of Chinese man-portable SAM systems began with the QW-1 (Vanguard-1), which went into production in the late 1990s. The design is based on the Soviet 9K310 Igla-1 (SA-16). Several versions have been developed over the past decade, including the QW-11, the QW-18 (armed with a two-band infrared seeker), and the QW-2 (Vanguard 2), which is a copy of the Igla-1 with an improved seeker. The latest modification is the QW-4, which is equipped with an IIR-type infrared seeker. The Chinese have also developed a heavier version with a larger missile, the QW-3, which is equipped with a semi-active laser seeker and has a range of up to 8 km. The QW-3 design has been used to develop several mobile versions mounted on truck and APC chassis; these have already gone into production. The QW-4 missiles are used in the TD-2000 self-propelled system, which has also gone into mass production. The NH-5 and QW-1 systems have been exported; all the latest systems are also being marketed to foreign customers.

China continues to make several types of towed anti-aircraft artillery, including the Type 90 35mm twin cannon (a licensed version of the Swiss Oerlikon Contraves GDF-002); the Type 87 25mm twin cannon (a version of the Soviet ZU-23-2 modified to use 25mm Oerlikon ammo); and the Type 80 23mm twin cannon (an unmodified copy of the ZU-23-2). The Type 87 and Type 80 designs have been used to develop several combined gun-missile systems; the missile component uses man-portable SAM missiles. Chinese exports in this segment include modified versions of Type 65/74 37mm twin AA cannons; the design can be traced back to the old Soviet V-47 twin cannon.

China's main self-propelled AA system uses four 25mm guns mounted on the chassis of Type 83 152mm self-propelled artillery. After producing a small batch of artillery-only Type 90 systems, the Chinese launched production of a combined gun-missile version, designated as Type 95 (PGZ-95), which is equipped with a QW-2 surface-to-air missile. China has recently developed a 35mm twin-gun Type 07 AA system mounted on a tank chassis; the design is similar to the old German Gepard. Another new mobile AA system is the LD-2000, which is essentially a Type 730 seven-gun 30mm shipborne AA artillery (a Chinese version of the Dutch Goalkeeper) mounted on a truck chassis. The Chinese are now developing a combined gun-missile version of the system equipped with the TY-90 guided missile. It appears that the LD-2000 will be supplied only to foreign customers.

Heavy armor

China's first tank, the Type 59, went into production at Plant No. 617 in Baotou (now part of CNGC) in 1958. It was a copy of the T-54 made under Soviet license and with Soviet assistance. In the early 1970s, China launched

Type 69; the design was a modified version of Type 59 and incorporated elements of a Soviet T-62 tank captured during an armed conflict with the Soviet Union on Damansky Island. Type 69 was followed by Types 79, 80, 85 and 88, which represented incremental improvements of the same basic design with more advanced Soviet or Western technologies. The gun used on these tanks was a Chinese copy of a British 105mm L7 system. Various versions of Type 59 were in production until the 1990s, and now form the core of the Chinese tank fleet. Large numbers have been exported to third world countries. In 1995, China launched a program to upgrade its Type 59 fleet to Type 59D specification, which includes a 105mm gun, a new fire control system and reactive armor. There is also another upgrade option, the Type 59P, developed by Poly Technologies for the export market. Type 85-IIAP tanks armed with a 125mm gun have been exported to Pakistan. First deliveries were made in 1991; a total of up to 400 units have been supplied so far. Another version offered to foreign customers is Type 85-III, which has reactive armor. Type 85-III design was used as a starting point in the development of Type 96 (ZTZ96, also known as Type 88C), which went into mass production at Plant No. 617 in 1997. Up to 2,500 units have been delivered to the PLA. The latest modification of the tank is Type 96G. It is now being marketed to foreign customers (including Bangladesh) as a cheaper option.

The latest generation of Chinese tanks began with the Type 90-II (MBT-2000); the design appears to be based on the Soviet T-72. Type 90-II was developed only for export; it is now being made in Pakistan under the local designation Al Khalid. But it was also used as a prototype for two other distinct third-generation Chinese tanks—Type 98 (ZTZ98) and Type 99 (ZTZ99). Only a few units of Type 98 have been made so far, while Type 99 went into mass production in Baotou in 2001 as the PLA's "first line" tank, along with Type 96. In 2007, China announced the Type 99A2 modification with better protection, and in 2008 it unveiled an even more advanced version, the Type 99G. The latest Type 99 modification was demonstrated in 2011. Up to 600 Type 99s have been built so far; none of them has been exported. The most advanced tanks China is now offering to its defense customers are the VT1 and VT1A modifications of the MBT-2000.

Despite clear progress in recent years, the Chinese are still lagging well behind world leaders in tank technology. Even their latest products are a rehash of older designs and solutions. The Chinese program to develop an advanced diesel engine for tanks is struggling. Type 99 tanks are equipped with MTU

MB871ka501 diesels made in China from German components. Limited supplies of these components appear to be one of the reasons why the Chinese are not making Type 99s in greater numbers. China also continues to import 6TD-2 engines and transmission blocks made by the Malyshev plant in Ukraine. These are used on the VT1A tanks offered to foreign customers.

Light armor

Over the past decade, China has launched mass production of a wide range of new-generation armored vehicles. As with most other Chinese weaponry, their design borrows heavily from Russian and Western technologies. For now, China's share of the world market for light armored vehicles is not large—but interest in new Chinese products in this segment is growing rapidly in developing countries. Amid the ongoing boom in the Chinese automotive industry, it is safe to expect that many new light armor models, including mine-resistant ambush protected vehicles, will be brought to market in the coming years.

China no longer makes Type 86 infantry fighting vehicles (WZ501, a copy of the Soviet BMP-1); Type 63 (WZ531), Type 77 (a copy of the BTR-50PK), Type 85 (WZ309) or WZ534 tracked armored personnel carriers; nor does it make WZ523 and Type 91 (ZBF91) 6x6 wheeled APCs. The new generation of Chinese light armor is represented by the 20 t Type 04 infantry fighting vehicle (ZBD04, previously designated as Type 97 or ZBD97). The vehicle's gun turret is Bakhcha-U, developed by Instrument Design Bureau in Tula and made under Russian license in China. Another new product is the 8 t Type 03 airborne combat vehicle (ZBD03, previously designated as ZLC2000). Type 03 and Type 04 were both developed with Russian assistance. China also makes Type 05 (ZBD05) amphibious armored fighting vehicles and Type 05 (ZTD05, Type 99) amphibious light tanks armed with a 105mm gun. The tracked APC segment is represented by Type 89 (YW535, ZSD89) and its modification, Type 90. The latter's chassis is used in the Type 02 (ZZC02) reconnaissance fighting vehicle and several other special-purpose vehicles. A modified version of Type 90, designated as VTP1, is offered to foreign customers.

The Type 90 (WZ551, ZSL90) 16 t 6x6 wheeled APC went into production in the early 1990s. In the mid-1990s, it was discontinued and replaced by Type 92 (WMZ551B, ZSL92), which remains in production to this day. The basic design is used in many special-purpose vehicles, including three-axle (WJ03B) and two-axle (Type 92B, WJ94) APCs in service with China's internal security forces. The Type 92I vehicle is also used as a chassis for a self-propelled anti-tank missile system and the Type 92 design was used in the new Type 02 (PTL02) 19 t fighting vehicle, which recently entered service with the PLA. The vehicle is equipped with a 105mm gun. The export version of it, marketed as the Assaulter, is also offered with an optional 120mm gun. In 2009, the PLA took first deliveries of the new-generation Type 09 8x8 wheeled APC (also designated as the VN1, ZBD09 and WZ0001). The chassis will be used in many other vehicles that are now in development. The family will include a fighting vehicle armed with a 120mm gun. There is also a 6x6 version, designated as the VN2.

The two-axle Chinese light armor now in production includes the ZFB05, a 6 t APC used by the police, and the VN-3, a 5 t reconnaissance fighting vehicle similar to the French Panhard VBL. In addition, China has developed the 7 t QL550 light armored vehicle, which is also similar to the Panhard VBL. All of these vehicles are offered to foreign customers.

Army trucks and off-road vehicles

China's automotive industry is booming, and the country has already become the world's largest market for cars and trucks. For that reason the Chinese auto industry can offer many militarized versions of civilian products. In fact, there are too many of these products to describe each one in detail.

The PLA's main light all-terrain vehicle is the BJ2020 (known previously as the BJ212), which is a copy of the Soviet UAZ-469. The more advanced BJ2022 version went into production in 2007, as did the 1.5 t EQ2050, a version of America's famous AM General HMMWV (Hummer H1). Nanjing Iveco Motor, a joint venture with Italy's Iveco, makes the 1.5 t NJ2045/NJ2046, a variation of the IVECO 40.10 model supplied to the PLA.

For a long time, China's main three-axle army truck was the Jiefang CA-30, a licensed version of the Soviet ZiL-157. The main two-axle army truck was the Jiefang CA-10, a copy of the ZiL-150. In the 1970s China also launched production of the three-axle EQ2081/2100 (a copy of the ZiL-131), and in the mid-1990s it started making the three-axle 3.5 t Dongfeng EQ2102 truck. In the 1990s, the older two-axle trucks were discontinued and replaced by the 1.5 t Dongfeng EQ2061, the 5 t Dongfeng EQ1093, and the 8 t Dongfeng EQ1141.

China's first heavy truck was the 7.5 t 6x6 Shaanqi SX2150, which went into production in 1969 and is still being made. The design combines the chassis of the Soviet Ural-375 with the cabin of the French Berliet GBU 16. In the mid-1990s, China launched production of the 7 t X2190, which is a clone of the Austrian Steyr 90. The 8 t three-axle CQ261, a licensed version of the French Berliet GBU and GCH, went into production in 1977. Production of the 8 t three-axle XC2030 (a copy of Mercedes-Benz 2026) began in 1986. Finally, in 1996 a joint venture with the Germans at Baotou launched production of the ND2629A (North-Benz 2629A) heavy three-axle truck and chassis.

Production of tank transporters and chassis for mobile missile systems began in the early 1980s using the HY473 chassis. In the 1990s, China launched production of the 12 t three-axle Steyr 24M heavy truck. In 1997 China's Wanshan Special Vehicle started making heavy multi-axle trucks designed by the Minsk Wheeled Chassis Plant for Chinese missile systems. One of the first products launched by the Chinese-Belarusian joint venture was a chassis based on the MAZ-543 design. At first it was assembled in China from Belarusian components. Later, the Chinese started making the localized version, the WS2400, which uses very few Belarusian parts.

In recent years, China's Wanshan Special Vehicle started manufacturing a whole range of heavy chassis, including eight-axle models similar to the MZKT-79221 (used for Russia's Topol-M ICBM). These will probably be used as mobile platforms for China's DF-31/31A ICBMs. The Chinese have also developed—with Belarussian assistance, apparently—the WS2500 10x10 chassis for new modifications of the DF-21 medium-range ballistic missile. There is also the WS2600 10x8 version, the WS2900 12x12 version, and the WS21050 14x12 version. All of them are based on MZKT designs, but according to Western sources they are equipped with Deutz diesel engines and ZF automatic transmissions imported from Germany. Wanshan Special Vehicle also makes the three-axle WS5251 and WS5252 chassis used in the new SH-series 122mm and 155mm self-propelled artillery and other systems.

Another Chinese supplier of heavy wheeled chassis is Taian Wuyue Special Vehicle, which makes the TA580/TAS5380 series 8x8 and 10x10 vehicles.

These are used as mobile platforms for the HQ-9 SAM system and the WM-80 MLR system.

Missile systems

China is the only official nuclear-weapon state that continues to develop medium-range missiles as delivery systems for nuclear weapons. The country is also developing and mass-producing non-nuclear tactical missiles; Taiwan is chief among the intended targets. Apart from ICBMs, the Chinese arsenal includes the old liquid-fuel medium-range DF-3 (up to 2,650 km range), DF-3A (2,800 km) and DF-4 (4,750 km) ballistic missiles. The new medium-range solid-fuel DF-21 and DF-21A/C have a range of 2,150 km and 2,500 km, respectively. The Chinese are developing new versions of the DF-21, including the anti-ship DF-21D missile. In the tactical segment, they have the solid-fuel DF-11 (M-11, 350 km), DF-11A (500 km) and DF-15 (M-9, 600 km) missiles.

In the past decade, China has developed a new generation of high-precision tactical missiles, including the P-12 (150-200 km range, similar to the Russian Iskander), the B-611 (150-200 km) and the B-611M (260 km). They also have several types of unguided missiles, such as the M-1B (100 km) and WS-1 (100-180 km), and their new guided versions, the WS-2 and WS-3 (200 km plus) and WS-2D (up to 400 km). Another guided missile system is the A200, with a range of up to 200 km. The Chinese have also developed a long-range guided version of missiles used in the WM-80 large-caliber MLR system, designated as WM-120 (120 km range).

In 2008, CASIC corporation demonstrated a mobile missile system armed with high-precision SY-400 missiles (400 km range). There have also been reports that the Chinese have developed the BP-12 satellite-guided missile for that system. Another interesting Chinese project is surface-to-surface systems based on surface-to-air missile designs. The M-7 missile is based on the HQ-2 SAM system and has a range of 150 km; the CF-2000 is based on the more advanced FT-2000 SAM and has a range of 180 km. According to US estimates, China produces more than 150 tactical missiles a year; they are deployed mostly across the strait from Taiwan.

China is one of the world's largest exporters of missiles in these segments, partly because its missile export policy is the least constrained by political considerations. It can be expected that given the new circumstances, China will soon end its exports of "destabilizing" missiles which fall under the restrictions of the Missile Technology Control Regime (including the 300 km range restriction), and will focus instead on exporting its most advanced missile systems such as the P-12, B-611, B-611M, A200, SY-400, WS-2 and WS-3.

Artillery systems

China is one of the world's biggest manufacturers of artillery. It has developed and launched production of a large variety of artillery systems, many of which are licensed or pirated copies of Soviet (Russian) or Western technology. Most of these systems are offered to foreign customers.

In the 1950s and 1960s, China launched production of many artillery towed systems under Soviet license. These included the D-44 85mm gun (Chinese designation Type 56), M-30 122mm howitzer (Type 54), D-74 122mm gun (Type 60), M-46 130mm gun (Type 59), D-1 152mm howitzer (Type 54) and D-20 152mm howitzer (Type 66). In later years, the Chinese copied and launched production of the D-30 122mm howitzer (Type 86 and its Type 96 modification). Only the Type 96 is still being manufactured and exported. The M-46 gun design was succeeded by the Type 83 152mm/54 gun with a range of 30 km for standard ammunition. The Chinese have also made a small batch of the domestically designed Type 86 152mm gun, with similar specifications. An upgraded version of the M-46 equipped with a 155mm/45 gun/howitzer barrel is exported under the GM-45 designation.

The new-generation Type 89 155mm/45 towed gun/howitzer (also designated as the W88/89, PLL01 and WA 021) went into production in 1989. The Chinese designers used the documentation for the GHN-45 gun/howitzer acquired from Austria's NORICUM. For its part, the GHN-45 is based on a design produced by Canada's Space Research Corporation. To their foreign customers the Chinese are also offering another version of the Type 89, designated as X52, with a 52-caliber barrel. Two versions of these systems are now being marketed by NORINCO, designated as AH1 (155mm/45) and AH2 (155mm/52). Yet another export version of this system is a 203mm/45 gun/howitzer, but not a single export contract has been signed for it so far.

The Type 83 122mm/45 towed howitzer is a domestically designed regimental artillery system. For their export customers the Chinese make the M-90

105mm mountain howitzer, which is a pirated copy of the Italian M56. In 2010, CNGC demonstrated the new AH4 155mm/39 towed howitzer; the design is similar to the Anglo-American M777.

The first Chinese self-propelled artillery was the Type 70 (WZ302) 122mm system, which was essentially a Type 54 howitzer fitted on top of a Type 63 tracked APC. Another early Chinese system was the Type 85 self-propelled turreted howitzer, which used a Type 83 122mm howitzer barrel mounted on a Type 85 wheeled APC chassis. Only a small number of Type 85 artillery systems were made. The design the Chinese chose for mass production was the Type 89 (PLZ89), a 20 t 122mm/32 self-propelled howitzer that was discontinued only recently. The Type 89 design was similar to the Soviet 2S1 Gvozdika self-propelled artillery. It used the artillery part of Type 86 howitzer in a turret mounted on a special floating tracked chassis.

Up to 500 units of the Type 89 self-propelled artillery had been made before it was discontinued, to be succeeded by a new generation of 122mm self-propelled systems. The first of them was demonstrated by CNGC in 2007. It was a self-propelled howitzer mounted on a 6x6 truck chassis with a longer barrel borrowed from the Type 96 howitzer. Another version of this system is the SH5, with a 105mm howitzer; both versions are offered to foreign customers. The elongated tipping unit of the Type 96 122mm howitzer is also used in two new self-propelled artillery systems: the SH3, which is mounted on a tracked chassis, and a self-propelled unit mounted on the new VN1 (ZBD09) 8x8 wheeled APC chassis. Meanwhile, Poly Technologies is marketing its own version of the Type 96 122mm howitzer mounted on a 6x6 truck chassis. The Chinese army has taken first deliveries of the new Type 07 (PLZ07) 122mm self-propelled howitzer, which uses a new tracked chassis similar to the Type 04 infantry fighting vehicle.

China's first 152mm self-propelled artillery was the 30 t Type 83 howitzer, a blatant rip-off ("semi-copy") of the Soviet 2S3 Akatsiya. Its barrel was borrowed from the Type 66 howitzer. A total of up to 500 units have been built. The Type 83 design was used to develop the Type 89 120mm self-propelled anti-tank gun (120 units built) and a 122mm MLR system also designated as Type 89. For the export market the Chinese then developed the PLZ45 self-propelled artillery using the same modified chassis (Type 321) and the barrel of the Type 89 155mm/45 gun/howitzer with a semi-automatic loading mechanism. The PLZ45 has been exported to Kuwait and Saudi Arabia;

a limited number of these systems have also been supplied to China's own army. China's latest self-propelled artillery is the Type 05 (PLZ05, PLZ52) 155mm/52 howitzer. It uses the Type 83 chassis and the artillery compartment of the Russian 2S19M1 Msta-S howitzer made under license in China. China's CNGC also offers the 22 t SH1 self-propelled howitzer, which uses a 6x6 truck chassis and a 155mm/52 gun based on the Type 89 design. According to some reports, the SH1 was designed specifically for export to Pakistan.

The main MLR systems now in service with the PLA are the old 107mm (Types 63 and 81) and 130mm (Types 63, 70, 82, 83 and 85) systems, as well as various clones of the Soviet BM-21 Grad, such as Types 81, 83, 84, 89 and 90. China continues to export 107mm and 122mm MLR systems and ammo. In the large-caliber segment, the Chinese have the 273mm Type 83, 284mm Type 74, and 305mm Type 79 systems, as well as the 253mm Type 81 and 87 mine-clearing rocket systems. Apart from the 122mm Type 90 system, Chinese companies are now marketing the 273mm eight-barrel WM-80 MLR system. Armenia bought four in 1999, and several have been supplied to North Korea. The Chinese are also offering the longer-range WM-120 version and a 425mm mine-clearing rocket system. Other Chinese products in this segment include the WS-1 MLR system (320mm caliber), WS-1B (320mm, bought by Turkey), WS-2 (400mm), WS-6 and WS-15 (122mm, Grad clones), and the WS-1D 252mm mine-clearing rocket system.

The most advanced Chinese MLR systems now in production are licensed versions of the Russian 300mm 9A52 Smerch, the A100 with 10 launch tubes, and the Type 03 (PHL03) with 12 launch tubes. The latter system has several export modifications, including the AR1 (eight launch tubes), AR1A (10 launch tubes), and AR2 (12 launch tubes). There is also a clone of the Smerch with eight launch tubes, designated as ANGEL-120.

The latest Chinese mortar is a licensed version of the Russian 2S23 Nona-SVK 120mm self-propelled mortar-gun system, designated as Type 05 (PLL05). Its artillery compartment is mounted on the WMZ551 6x6 wheeled chassis. Chinese companies also make and export regular mortars of 60mm caliber (Types 63 and 90), 81mm (Types 87, 91 and 99), 82mm (Types 53, 67 and 84), 100mm (Type 71) and 120mm (Types 53, 55 and 86). Most of them are copies or modifications of early Soviet or modern Western designs. Export versions include a pirated copy of the Soviet 82mm 2B9 Vasilyok (W99) automatic mortar, including a self-propelled version.

China is also marketing the GP1 155mm guided artillery projectile with a semi-active laser seeker (the Chinese version of the Russian Krasnopol-M ammo) and the GP4 120mm guided mortar bomb.

Anti-tank weapons

For a long time, China's main anti-tank missile system was the HJ-73, a copy of the Soviet 9K14 Malyutka (AT-3). It went into production in 1979. The last versions of the system the Chinese developed before discontinuing the whole family were the HJ-73B (with a semi-automatic targeting system used in the HJ-8 system) and the HJ-73C (with a new semi-automatic system and a tandem charge). In 1987, China launched production of the HJ-8 mobile system; the design is similar to the American TOW. The system has a range of 3,000 m and a semi-automatic wire guidance system. Several modifications have been developed over the years. The HJ-8E version, which entered production in the late 1990s, has a new charge and a longer (4,000 m) range. The system can also be used with the Z-9 helicopter, and there is also a lighter man-portable version, the HJ-8L.

The HJ-9, which entered production in the late 1990s, is an incremental improvement of the HJ-8, with a 5,000 m range and a semi-active laser guidance system. China has reportedly developed the HJ-9A version with a millimeter-band radar seeker, and the HJ-9B with a laser ray guidance system. For their latest Z-10 attack helicopter, the Chinese have developed the long-range (up to 8 km) HJ-10 (AKD-10) anti-tank missile system equipped with a semi-active laser seeker. China has also announced the development of domestic guided 105mm and 125mm tank gun missile projectiles with a laser guidance system. Their design is likely based on the Russian Bastion, Refleks and Basnya missiles, all of which are made under Russian license in China.

Chinese rocket-propelled grenades now in production include the Type 69 (a copy of the Soviet RPG-7) and its improved version, the Type 4. The Chinese have developed a large number of projectiles for these RPGs. They continue marketing the obsolete Type 70-I 62mm RPG, which was developed domestically. There is a more recent system, the 120mm Type 98 (PF98, Queen Bee). The standard Chinese single-use RPG system is the 85mm Type 89 (PF89), and the PF98-1 version with a thermobaric (PT) charge. NORINCO continues to market to foreign customers the Type 78 82mm recoilless gun, which is a copy of the Soviet B-10.

Small arms and infantry weapons

China is one of the world's leading suppliers of small arms and infantry weapons. Chinese weaponry and ammo in this segment are affordable, simple and reliable, and Beijing is willing to sell them without any particular political preconditions.

The Chinese defense industry makes a broad range of small arms, from copies of old Soviet designs to advanced domestic systems. Still, the Chinese weapons that are in the greatest demand on the market are first-generation systems that went into production in the 1950s and 1960s under Soviet license, or as unlicensed copies of Soviet weaponry. The most popular Chinese exports in this segment are the Type 56 7.62mm assault rifle (a copy of the Kalashnikov design dating back to 1947), the Type 63 assault rifle, and the Type 74 light machine gun. The latter two systems are both based on the Type 56 design. The "Chinese Kalashnikov" has become a popular brand with many armies and armed groups in the developing world.

The first-generation Chinese weapons also include: the Type 56 7.62 semi-automatic carbine (a copy of the SKS); the Type 54 7.62mm pistol (a copy of the TT; there is also a 9mm version called Model 213); the domestically designed Types 64 and 77 7.62mm pistols (there is a 9mm version of Type 77); the Type 59 9mm pistol (a copy of the PM); the Type 64 7.62mm submachine gun (a descendant of the PPS); the Types 79 and 85 7.62mm sniper rifles (copies of the SVD); the Type 67 7.62mm machine gun (a copy of the PK) and its Type 80 variant; the Type 54 12.7mm machine gun (a copy of the DShKM) and its variant, the Type 59; and the Type 56 14.5mm machine gun (a copy of the KPV) and its Type 58 and 75 variants.

In the 1980s, the Chinese launched a new generation of small arms based on a redesign of the Kalashnikov family, including the 7.62mm Type 81 assault rifle and Type 81 machine gun. They also designed a 7.62mm Bullpup-configuration Type 86 assault rifle; Types 79 and 85 7.62mm submachine guns, and 12.7mm Type 77, 86, 89 (QJZ89) and W85 machine guns.

In the late 1980s, the PLA took first deliveries of a radically new generation of small arms, which made use of the original Chinese Type 87 (DBP87) 5.8mm (5.8x42) ammo; the newer and more popular variant of that ammo is Type 95 (DBP95). Weapons using this new ammunition make up the bulk of the small arms now in service with the PLA. These include the Type 95 (QBZ95) 5.8mm Bullpup-configuration automatic rifle and the Type 95 (QBB95) light machine gun. In 2010, it was reported that the Chinese had developed a new Type 95 rifle known as Model G; the weapon uses improved 5.8mm ammo. Chinese engineers used the basic Kalashnikov assault rifle design to develop a new Type 03 (QBZ03) 5.8mm assault rifle, although it is not clear how many of those have been delivered to the PLA. A heavier version of the 5.8x42 ammo, known as Type 88 (DBP88) is used in the Type 88 (QBU88) 5.8mm sniper rifle and Type 88 (QJY88) machine gun. The shorter 5.8x21 Type 92 ammo (DAP92) is used in the Type 92 pistol (QSZ92, there is also a 9mm version) and Type 06 (QCW06) pistol, as well as the Type 05 submachine gun (QCW05). For the 9mm ammo the Chinese have developed Type 06 (LS06) and JS submachine guns.

For export, the Chinese manufacture automatic weapons that use the NATO-standard 5.56mm M193 ammo. These include Type 97 (QBZ95, a version of Type 95), Type 06 (a version of Kalashnikov Type 56) and CQ (M311, a copy of the American M16/M4) automatic rifles, as well as Type 97 sniper rifles, which are a variant of the Type 88. The JG 7.62mm sniper rifle is also made for export. There is no reliable information about Chinese exports of 5.8mm small arms, and exports of weapons using 5.56mm ammo appear to be quite limited.

The latest Chinese large-caliber weapons now in production are the Type 02 (QJG02) 14.5mm machine gun and Type 99 (M99), Type 06 (M99B), AMR-2 and JS 05 12.7mm sniper rifles. The LM06 12.7mm machine gun, which is a copy of the American M2HB, is offered to foreign customers.

China's main anti-infantry grenade launcher is Type 87 (QLZ87, W87), which uses original Chinese 35mm projectiles. There is a handheld and mounted version, as well as the improved QLZ06 (QLZ87B) variant. The Chinese also make a copy of the Soviet AGS-17 30mm automatic grenade launcher. The LG02 40mm rifle-attached grenade launcher can be used with Type 95 and 97 rifles; there is also the LS03 40mm export version. The Chinese Type 97 handheld infantry flame-throwers are a licensed version of the Russian RPO-A Shmel. Finally, the Chinese make a broad range of rifle grenades.

3.6 Conclusion

Since the late 1990s, the Chinese defense industry has radically refreshed its entire product range. In most areas it has substantially narrowed the gap with the leading Russian and Western suppliers, and in some cases closed that gap altogether. Nevertheless, China's progress in defense technology still relies heavily on borrowing and adapting foreign designs and solutions. The country has managed to minimize imports of finished weapons platforms such as aircraft and ships, but it still depends on imports of key components, including engines, radars, and targeting systems. China is making energetic efforts to end that dependence, but the quality of domestic Chinese designs still leaves much to be desired. Meanwhile, imported technologies often take the Chinese engineers so long to assimilate that they become obsolete by the time the process is completed. However, China has become competitive in such segments of the defense market as small arms, heavy and light armor, artillery systems and army trucks and chassis. And the new Chinese J-10 and FC-1 fighter jets have good export potential—but any hypothetical Chinese expansion in this segment is being held back by the need to import the engines for both aircraft from Russia.

Economically, Chinese defense contractors are not dependent on arms exports. Most of their revenues are generated by the civilian side of their business. Besides, any theoretically achievable export figures are dwarfed by China's own arms procurement programs, which are growing at a steady pace. Beijing does not need arms exports to keep its defense industry afloat. It views these exports mainly as an instrument to bolster its foreign policy clout and international prestige. As a result, China is prepared to be very flexible when negotiating the financial terms of arms contracts. It often links these terms to unrelated areas of trade and economic cooperation, or uses them as a bargaining chip to secure political concessions. In some cases it even supplies weapons as gifts of aid. Conditions for the expansion of Chinese arms exports are the most favorable in poorly developed countries that depend on exports of raw materials, pursue anti-Western policies or simply prefer to hedge their bets. Another potential target for Chinese expansion is economies that are gradually developing a dependence on the Chinese market or Chinese investment.

At the same time, China's transformation into a global superpower has some obvious benefits for Russian arms exporters. China's growing military might and arms exports will force countries such as India and Vietnam to ramp up their own defense spending, and the size of their defense markets will increase. In addition, growing global rivalry between China and the US may make buying weapons from Beijing politically inexpedient for those countries that have reason to view China as a potential national security threat.

Notes

1. Data taken from a report by China Lianhe Credit Ratings on assigning a credit rating to CNGS corporate bonds, April 30, 2010. <http://bond.hexun.com/upload/zhongguobingqi.pdf>.

2. See "CNGC: Military Establishes Base of Civilian Products" [军品立位，民品立业], Jan. 28, 2002. china.com.cn. <http://big5.china.com.cn/chinese/junshi/101863.htm>.

3. See "Guidelines on the participation of non-state-owned companies in the development of the defense industry" [关于非公有制经济参与国防科技工业建设的指导意见], Oct. 30, 2008. <http://jmjhs.miit.gov.cn/n11293472/n11295193/n11298643/ 11618138.html>.

4. Wang Shanshan, Fu Tao and Yu Ning. "SOEs Ordered to Check Out of Hotel Sector" [央企酒店大重组"较劲"], March 15, 2010. Caixin Online. <http://magazine.caixin.com/2010-03-14/100126180.html>

5. Information from a SASAC press conference, March 18, 2010. <http://www.sasac.gov.cn/n1180/n1566/n259730/n264168/7019501.html>.

6. Mulvenon J. and R. Tyroler-Cooper. "China's Defense Industry on the Path of Reform." Prepared for the US-China Economic and Security Review Commission, October 2009, p. 12.

7. Ibid. p. 8.

8. Medeiros E. *A New Direction for China's Defense Industry.* Rand Corporation, 2005, pp. 34-35.

9. Ibid. p. 48.

10. Mulvenon and Tyroler-Cooper, op. cit., p.10.

11. From *China Legal Evening News* [中国法制晚报], Nov. 8, 2008.

12. Medeiros, op. cit., pp. 62-63.

13. COMAC Corporate Website. <http://www.comac.cc/gk/gsjj/>.

14. From China's *First Financial Daily* [第一财经日报], June 27, 2008.

15. See "China launches program to hire foreign engine specialists." *RIA Novosti*, May 31, 2010.

16. See "Key Developments, AVIC International Holding Ltd." <u>Reuters</u>. <http://uk.reuters.com/business/quotes/keyDevelopments?symbol=0232.HK&pn=1>.

17. See "AVIC General Aircraft Company sets goal to capture one-third of the global market by 2025," [中航通飞计划2025年交付量占全球1/3]. *Economic Observer* [经济观察报], July 6, 2009. <http://aero.huanqiu.com/ga/2009-07/506589.html>.

18. See "Avicopter established...," [中航工业直升机公司成立 计划明年整体上市]. *Economic Observer* [经济观察报], Feb. 27, 2009. <http://finance.sina.com.cn/chanjing/b/20090227/12065910612.shtml>.

19. From *21st Century Business Herald* [21世纪经济报道], Nov. 23, 2006.

20. China North Industries Group Corporation is the official English name of the parent company. The Chinese name is 中国兵器工业集团公司, which means "China Ordnance Industry Group Corporation." This is the name used by the official Chinese press in Russian-language materials.

21. <u>CSGC Corporate Website</u>. <http://www.csgc.com.cn/n27/n45/index.html>.

22. See "Interview with China North Industries Group Development Strategy Committee member Zhang," [专访中国兵器工业集团发展委员会员张忠], Nov. 28, 2008. <http://www.jichuang.net/jc_v3/jc_news/Jc_news_show.asp?id=61652&sort=22>.

23. <u>Inner Mongolia First Machinery Group Corporation Corporate Website</u>. <http://www.nmgyj.com>.

24. The company in Baotou is the only supplier of main battle tanks to the PLA. It produces two models, Type 96 (and its various modifications) and Type 99G. By most estimates, the annual output of Type 96 is up to 200 units, and 30-40 units of Type 99G.

25. See "The mysterious rise of Inner Mongolia as a 'kingdom of cars,'" [内蒙古"重车王国"崛起之谜], Dec. 28, 2007. <u>North News</u>. <http://www.northnews.cn/2007/1228/119960.shtml>.

26. <u>CNGC Corporate Website</u>. <http://www.cngc.com.cn/intro1.aspx>.

27. See "China North Industries Group Corporation 2009 report card," [中国兵器工业集团公司2009 年交出精丽的成绩单]. *Xinhua News Agency*, Jan. 19, 2010.

28. <u>Huajin Chemicals Corporate Website</u>. <http://www.huajinchem.com/>.

29. <u>Zhenhua Corporate Website</u>. <http://www.zhenhuaoil.com/en-rlzy.htm>.

30. See "Chinese weapons maker acquires oil import license." *RIA Novosti*, June 30, 2010.

31. <u>Zhenhua Corporate Website</u>. <http://www.zhenhuaoil.com/en-yw-my.htm>.

32. For more details about AVIC reform see Kashin V., "Chinese aerospace industry undergoing radical reform." *Eksport vooruzheniy* [Arms Exports], No. 3, 2010.

33. See "China North Industries Group Corporation 2009 report card," [中国兵器工业集团公司2009 年交出精丽的成绩单]. *Xinhua News Agency*, Jan. 19, 2010.

34. CALT Corporate Website. <http://www.calt.com/bygk/index.html>.

35. CAST Corporate Website. <http://www.cast.cn/about/cjyfz.html>.

36. Medeiros, op. cit., p. 56.

37. Ibid. p. 57.

38. CASIC Corporate Website. <http://www.casic.com.cn/n101/index.html>.

39. CSSG Corporate Website. <http://www.cssg.com.cn/Item/35.aspx>.

40. CGWIC Corporate Website. <http://cn.cgwic.com>.

41. CSIC Corporate Website. <http://www.csic.com.cn/jtjtfc/zgjtjj>.

42. See *Xinhua News Agency*, Jan. 25, 2010

43. CSSC Corporate Website. <http://www.cssc.net.cn/en/component_general_situation/>.

44. See *Xinhua News Agency*, March 30, 2010.

45. CETC Corporate Website. <http://english.cetc.com.cn/business.html>.

46. CETC Corporate Website. <http://www.cetc.com.cn>.

47. The list of subsidiary companies is available at CETC's corporate website <http://www.cetc.com.cn/Article_List.aspx?columnID=10>.

48. Mulvenon and Tyroler-Cooper, op. cit., pp. 24-25.

49. See "CETC International integrates operations," July 12, 2010. china.org.cn. <http://www.china.org.cn>.

50. CEIEC Corporate Website. <http://www.ceiec.com.cn>.

51. Interview with Wan Tianmin, the chief designer of the ZBD-04 infantry fighting vehicle, in the journal 坦克装甲车辆 [Tanks and Armor], No. 3, 2010.

52. See "Chinese purchases of Russian engines," [Китайские закупки российских двигателей], July 4, 2011. Russian Arms [ОРУЖИЕ РОССИИ]. <http://www.arms-expo.ru/049051124050052048056051.html>.

53. Karnozov V. "Big New Chinese Order for Russian Fighter Engines," Oct. 3, 2011. Aviation International News Online. < http://www.ainonline.com/aviation-news/ain-defense-perspective/2011-10-03/big-new-chinese-order-russian-fighter-engines>.

54. See "Continental Motors Sold to China AVIC." *Aviation Week*, Dec. 7, 2010.

55. AVIC Corporate Website. <http://www.avic.com.cn>.

56. Corporate address: 25 Shuntunlou Street, Shunyi, Beijing, [北京市顺义区顺通路25号], corporate registration number 1000000042774.

57. See "Acquisition summary report...," [航空动力：收购报告书摘要], June 28, 2011. Stock Star. <http://stock.stockstar.com/JI2011062800000347_1.shtml>.

58. For information on the transfer of Xian Aero Engine PLC assets to AVIC Aero-Engines, see "Acquisition summary report...," [航空动力：收购报告书摘要], June 28, 2011. Stock Star. <http://stock.stockstar.com/JI2011062800000347_3.shtml>.

59. A publicly traded subsidiary of Xian Aero Engine Group Company, see XAEC Corporate Website. <http://www.xaec.com>.

60. See "AVIC Aero-Engines conducting major consolidation of assets, preparing to become a single platform," [中航发动机资产大整合 控股公司成整合平台]. *21st Century Business Herald* [21世纪经济报道], June 28, 2011.

61. See "Acquisition summary report...," [航空动力：收购报告书摘要], June 28, 2011. Stock Star. <http://stock.stockstar.com/JI2011062800000347_3.shtml>.

62. A large manufacturer of aircraft engine components, including nozzle control units and fuel injection control systems. See Beijing Changkong Machinery Company Corporate Website. <www.bkck.com.cn>.

63. Data from a report by AVIC Aero-Engine Controls Co. on changes in the list of shareholders, June 23, 2011. Shenzhen Stock Exchange. <http://disclosure.szse.cn/m/finalpage/2011-06-24/59590669.PDF>.

64. Ibid.

65. In September 2011, the Chinese name of the company was changed from 中航商用飞机发动机有限责任公司 (AVIC Limited Liability Company for Commercial Airplane Engines) to 中航商用航空发动机有限责任公司 (...for Commercial Aircraft Engines), which could be an indirect indication that the company's product range will also include helicopter engines.

66. ACAE Corporate Website. <http://www.acae.com.cn>.

67. Ibid.

68. Ibid.

69. See "Wang Lin: Chinese-made aircraft engines to be available by 2020," [王之林:国产大飞机发动机拟2020年交付用户], April 4, 2011. China Daily. <http://www.chinadaily.com.cn/hqcj/2011-04-04/content_12273074.htm>.

70. ACAE Corporate Website. <http://www.acae.com.cn>.

71. Ibid.
72. Ibid.
73. Ibid.
74. AVIC Harbin Dongan Engine (Group) Corporation Ltd. [中航工业哈尔滨东安 发动机（集团）公司] Corporate Website. <http://www.dongangroup.cn>.
75. Ibid.
76. Ibid.
77. Ibid.
78. Ibid.
79. Shenyang Liming Aero-Engine Group Corporation Corporate Website. <http://www.lmaeg.com>.
80. Information from the Jobs section. Northwestern Polytechnical University Website. <http://job.nwpu.edu.cn>.
81. Ibid.
82. For more background on SAERI, see <http://wenku.baidu.com/view/29b68417866fb84ae45c8dc6.html#>.
83. CNSAIC Corporate Website. <http://www.cnsaic.com>.
84. SASAC Corporate Website. <http://www.sasac.gov.cn>.
85. CNSAIC Corporate Website. <http://www.cnsaic.com>.
86. Ibid.
87. Ibid.
88. Ibid.
89. See "WZ-8D turboshaft aircraft engine," [涡轴８Ｄ型航空发动机], Nov. 9, 2006. Guangzhou Government. <http://www.guangzhou.gov.cn/special/2006/node_1409/node_1411/2006/11/09/1163058676138946.shtml>.
90. See *Jane's Defense Weekly*, Feb. 17, 2010.
91. XAEC Corporate Website. <http://www.xaec.com>.
92. See "Xi'an Aero Engine Company Ltd. 2011 Semi-Annual Report," [西安航空动力股份有限公司 2011 年半年度报告], July 27, 2011. <http://static.sse.com.cn/cs/zhs/scfw/gg/ssgs/2011-07-29/600893_2011_z.pdf>.
93. See "Xi'an Aero Engine Company Ltd. 2010 Annual Report Summary," [西安航空动力股份有限公司2010年年度报告摘要], March 14, 2011. Sohu.com. <http://stock.sohu.com/20110314/n279818130.shtml>.
94. XAEC Corporate Website. <http://www.xaec.com>.

95. Ibid.
96. See Galaxy Securities Company Ltd. report, Aug. 23, 2010. <http://file.finance.sina.com.cn/211.154.219.97:9494/MRGG/CNSESH_STOCK/2010/2010-8/2010-08-23/597273.PDF>.
97. See "Chengdu Engine Group and its WS18," [成都发动机(集团)有限公司与其WS18], Nov. 6, 2009. <http://www.fyjs.cn/viewarticle.php?id=216057>.
98. Ibid.
99. This section uses data accumulated by the Center for Analysis of Strategic and Technologies (including the book *Russian Military-Technical Cooperation with Foreign Countries: Markets Analysis*, Moscow, Nauka, 2008), information from Jane's publications (www.janes.com), as well as reports from world.guns.ru, www.sinodefence.com, cnair.top81.cn and other sources (listed separately).
100. Xi'an Aircraft Industry Company Corporate Website. <http://www.xiancn.com/gb/news/2008-05/09/content_1451831.htm>.

Index

Note: Page numbers in italics indicate material in tables.

imports to China after, 70–71
imposition of, 50, 70–71, 79
lifting of, hope for, 71, 73–74
arms market
China as exporter, 49, 71, 75–76, 77–137. *See also* exports by China
China as importer, 49–75. *See also* imports by China
China's position on, 49–51, 136–137
China's presence on (1992–2010), *82*
latest trends on, 71–76
army, Chinese. *See also specific weapons and equipment*
CNGC as supplier of, 10–13
General Armament Department of, 2, 4, 5–6
Arsenal Design Bureau (Ukraine), 117
artillery systems
Chinese export of, 80, 130–133
to Bangladesh, 91
to Bolivia, 92
to Indonesia, 94
to Kuwait, 80–81, 95
to Mexico, 95
to Pakistan, 80, 81, 86
to Rwanda, 97
to Saudi Arabia, 97
Chinese import of, 54, *60*
Chinese production of, 11, 13, 130–133
Iranian production under Chinese license, 87
shipborne. *See* naval weapons systems
ASN–7 unmanned aerial vehicle (China), 115
ASN–9 unmanned aerial vehicle (China), 115
ASN–15 unmanned aerial vehicle (China), 115
ASN–229A unmanned aerial vehicle (China), 116
Aspide missile (Italy), 68, *70,* 117–118, 123
Assaulter armored vehicle (China), 127
assault rifles, Chinese, 134–135
Austria, artillery technology from, 130
automotive industry, Chinese, 10, 126, 127–129
Aviation Industry Corporation of China (AVIC), *1,* 6–10

assets of, 6
Bolivian contract with, 92
Burmese contract with, 88
civilian business of, 6–7
engine production of, 9, 35–48
flotation strategy of, 10
foreign trading division of, 2–3
Pakistani contracts with, 85
restructuring (reform) of, 4, 6, 7
subsidiaries of, 7–8, 7–10
Aviation Industry Corporation I of China (AVIC I), 6
Aviation Industry Corporation II of China (AVIC II), 6
AVIC. *See* Aviation Industry Corporation of China
AVIC Aero-Engine Company, 7, 9, 35, 38–39
AVIC Aero-Engine Controls (AAEC), 38–39
AVIC Aircraft, 7, 8
AVIC Assets Management Division, 7, 9
AVIC Automobile, *8,* 10
AVIC Chengdu Engine Group (CEG, Plant No. 420), 44–45
AVIC Commercial Aircraft Engine Company (ACAE), 7, 9, 37, 38, 39–41
AVIC Construction Projects Company, *8*
AVIC Defense Industry Division, 7, 9
AVIC Economic Research Institute, *8*
AVIC Flight Test Establishment, *8*
AVIC Fundamental Research Technology Institute, *8*
AVIC General Aircraft Company, 7, 8, 10
AVIC Helicopter Company (Avicopter), 7, 8–9, 10, 43
Avichina Industry & Technology Company Ltd., 7, 9
AVIC International, 3, 7, 9
AVIC Investment, *8*
Avicopter, 7, 8–9, 10, 43
AVIC Systems, 7, 9
AVIC Xian Aeroengine (Group) Ltd. (XAEC, Plant No. 430), 38, 43–44, 45
Azerbaijan, Chinese exports to, 75, 91, 107

B

B–611/611M ballistic missile (China), 129–130

BA–7 missile (China), 118

Bakhcha–U gun turret (Russia to China), 54, *61*, 126

ballistic missiles
Chinese export of, 85, 129–130
Chinese production of, 13–15, 24, 129–130

Bangladesh, Chinese exports to, 49, 91–92
aircraft, 75, 91–92, 103, 107
ships, 92, 120, 121

Baoding rotor plant, 9

Basnya missiles (Russia to China), 54, *61*, 133

Bastion missiles (Russia to China), 54, *61*, 133

Beihai Shipbuilding Heavy Industry Company, 22, 26

Beijing Changkong Machinery Company, 38

Beijing Institute of Aerodynamics, 116

Beijing Raise Science, 9

Beijing University of Aeronautics and Astronautics (BUAA), 115, 116

Beijing Wisewell Avionics Science and Technology Company, 116

Belarus, missile chassis from, 128

BJ2020 all-terrain vehicle (China), 126

BJ2022 all-terrain vehicle (China), 126

Blue Eagle unmanned aerial vehicle (China), 116

Bohai Heavy Industry Company, 22, 24

Bohai shipyards, 16, 22, 24, 25, 26

Bolivia, Chinese exports to, 75, 92

BP–12 satellite-guided missile (China), 129

Britain. *See* United Kingdom

BUAA. *See* Beijing University of Aeronautics and Astronautics

Burma, Chinese exports to, 80, 82, 87–88
aircraft, 75, 88, 103
ships, 88, 120

BZK–005 unmanned aerial vehicle (China), 116

BZK–7 006 unmanned aerial vehicle (China), 115

BZK–009 unmanned aerial vehicle (China), 116

C

C–601 missiles (China), 118

C–611 missiles (China), 118

C–701 missile (China), 118

C–704 missile (China), 118

C–705 missile (China), 118

C–801 missiles (China), 80, 118–119, 121

C–802 missiles (China), 80, 118–119, 121

C–803 missiles (China), 118–119

C919 jetliner (China), 6, 39–40

CALT. *See* China Academy of Space Launch Technology

Cambodia, Chinese exports to, 78, 92, 120, 121

Canada
aircraft engines from, 110, 112–114
artillery technology from, 130

CASC. *See* China Aerospace Corporation

CASET. *See* China Academy of Space Electronics Technology

CASIC. *See* China Aerospace Science and Industry Corporation

CAST. *See* China Academy of Space Technology

CATIC. *See* China National Aero-Technology Import & Export Corporation

CCETDI. *See* China Changfeng Electromechanical Technology Design Institute

CEG. *See* Chengdu Engine Group

CEIEC, 17

Central Military Council, China (CMC), 2, 34

CETC. *See* China Electronics Technology Corporation

CF–2000 missile (China), 129

CFM International, 39, 46, 74, 111

CGWIC. *See* China Great Wall Industry Corporation

CH–3 unmanned mini-helicopter (China), 116

Chad, Chinese exports to, 93

CHAIG. *See* Changhe Aircraft Industry

China Shipbuilding Trading Company
(CSTC), 16, 19
China South Industries Group
Corporation (CSGC), *1*, 10
China State Shipbuilding Corporation
(CSSC), *1*, 15–16, 18–19, 26–33
civilian business of, 16, 19
foreign trade division of, 19
Guangzhou operations of, 19, 26–27,
31–32
Jiujiang operations of, 26, 32–33
military programs of, 16, 19
restructuring (reform) of, 15–16, 18, 19
sales volume of, 19
Shanghai operations of, 19, 26, 27–31
subsidiaries of, 26–33
"Chinese Kalashnikov," 134
Chongming Island shipyards, 19, 30
Chongqing Shipbuilding Industry
Company, 18, 22, 25
CIS. *See* Commonwealth of Independent
States
CJ–5 trainer (China), 108
CJ–6 trainer (China), 108
CMC. *See* Central Military Council
CNECC. *See* China Nuclear Engineering
& Construction Corporation
CNGC. *See* China North Industries
Group Corporation
CNGC Huajin Chemicals, 13
CNNC. *See* China National Nuclear
Corporation
CNSAIC. *See* China National South
Aviation Industry Company
COIG. *See* China Ordnance Industry
Group
COMAC. *See* Commercial Aircraft
Corporation of China
combat aircraft, Chinese, 102–108. *See also
specific aircraft*
J–7 fighter, 103–104
J–8 fighter, 104
J–10 multirole fighter, 104–105
combat trainers
Chinese export of, 108–110
to Angola, 90
to Bolivia, 92
to Burma, 88

to Democratic Republic of
Congo, 93
to Egypt, 89
to Ghana, 93
to Kenya, 94
to Morocco, 95–96
to Namibia, 96
to Pakistan, 80, 84
to Sri Lanka, 98
to Sudan, 89
to Tanzania, 99
to Venezuela, 101
to Zambia, 101
to Zimbabwe, 102
Chinese import of, 52, *55*
Chinese production of, 108–110
engines for, 37, 109
joint China–Pakistan program on, 84
joint China–Ukraine program on, 100
Commercial Aircraft Corporation of
China (COMAC), 6, 39–40
commercial period of Chinese exports
(1980–1988), 77, 78
Commission of Science, Technology
and Industry for National Defense
(COSTIND), 4, 5
Commonwealth of Independent States
(CIS)
China's acquisition of technology from,
80
Chinese competition with, 76
Chinese exports to, 75, 82
Congo, Democratic Republic of. *See*
Democratic Republic of the Congo
Continental Motors, 37
corporations. *See* defense corporations;
specific corporations
COSTIND. *See* Commission of Science,
Technology and Industry for
National Defense
CPMIEC. *See* China Precision Machinery
Import & Export Corporation
Creusot–Loire T100C artillery system
(France to China), 68, *69*
Crotale SAM system (France to China),
68, *69*, 86, 123
CSGC. *See* China South Industries Group
Corporation

CSIC. *See* China Shipbuilding Industry Corporation
CSSC. *See* China State Shipbuilding Corporation
CSSG. *See* China Sanjiang Space Industry Group Company
CSTC. *See* China Shipbuilding Trading Company

D

D–4 unmanned aerial vehicle (China), 115
Dalian Marine Diesel Works, 24
Dalian New Shipbuilding Heavy Industries Corporation, 22, 23
Dalian Shipbuilding Industry Company, 18–19, 22–24
Dalian Shipyard, 16, 18–19, 21, 22–24
Dassault, 74
Datang Telecom Technology Company, 17
DCK–007 unmanned aerial vehicle (China), 116
defense corporations, Chinese, *1,* 1–3. *See also specific corporations*
  civilian (commercial) business of, 2
  distinguishing features of, 2
  economic conditions for, 136
  foreign trading divisions of, 2–3
  government control of, 3–6
  management performance of, 2
  mergers of, 4
  military contracts of, 2
  non-state-owned, 3
  restructuring of, 3, 4
  specialized intermediaries of, 3
  state ownership of, 1
  subsidiaries of, 2, 3
defense industry, Chinese. *See also specific entries*
  as competitor to Russia, 76, 136
  development strategy for (SASTIND), 4–5
  general state of, 1–3
  government control of, 3–6
  growth of, 74, 136–137
  institutional landscape of, 6–17
  key defense corporations in, *1,* 1–3
  nationalism in, 73

DEMC. *See* Harbin Dongan Engine Manufacturing Company
Democratic Republic of the Congo, Chinese exports to, 75, 93, 110
Department for the Promotion of Integration of Military and Civilian Programs, 5
development strategy, Chinese, 4–5
DF–3A ballistic missile (China), 129
DF–3 ballistic missile (China), 129
DF–4 ballistic missile (China), 129
DF–11 ballistic missile (China), 129
DF–15 ballistic missile (China), 129
DF–21 ballistic missile (China), 129
domestic security equipment, Chinese import of, 66, 71, 75
Dongfeng EQ1093 truck (China), 126–127
Dongfeng EQ1141 truck (China), 126–127
Dongfeng EQ2061 truck (China), 126–127
Dongfeng EQ2102 truck (China), 126
DUF–2 micro-unmanned aerial vehicle (China), 116

E

East Timor, Chinese exports to, 120
Ecuador, Chinese exports to, 75, 79, 81, 93
Ecureuil AS350 helicopter (France), 67–68, *69,* 112, 114
Egypt
  Chinese exports to, 49, 82, 88–89
    aircraft, 89, 103, 107, 110
    ships, 120
  Soviet aircraft acquired from, 46
electronics, Chinese, 16–17
embargo. *See* arms embargo
engineering training, in China, 5
engines, aircraft. *See* aircraft engines
EQ series trucks (China), 126–127
Eritrea, Chinese exports to, 75
Eurocopter, 43, 68, 112–114
Europe
  Chinese imports from, 65, 67–71
    aircraft engines, 36, 45, *70,* 74, 108, 112–114

air defense systems, 68–69,
69–70, 71
after arms embargo, 50, 70–71,
74, 79
before arms embargo, 67–69
helicopters, 67–68, *69*, 112–114
lifting of embargo, hope for, 71,
73–74
shipborne weapons systems, 68,
69
technology transfer as priority, 74
US pressure against, 74
Chinese weapons systems relying on
technology from, *69–70*
exports by China, 77–137. *See also specific
exports*
1949 to 1992, 77–78
1992–, 79–82
to allies, 82
annual sales (1992–2010), *82*
anti-Western or anti-American policies
in, 76
China as net exporter, 49, 71, 75–76, 78
commercial period of (1980–1988),
77, 78
competition with Russia, 76, 136
cost as factor in, 76, 83
dollar value of
1992–2000, 80, *82*
2001–2005, 81, *82*
2006–2010, *82*
expansion of
priority markets for, 81–82
rapid period of (2005–2010), 81
geopolitical period of (1970s), 77, 78
ideological period of (1950s–1960s),
77–78
importers of. *See also specific countries*
biggest (1992–), 82–90
Iran–Iraq War as stimulus for, 49, 77,
78
to pariah states, 82
political diversification through, 83
statistics, reports and estimates of, 82
technological limitations of, 79
exports to China. *See* imports by China;
specific exporting countries

F

F–7. *See* J–7 fighter (China)
F–22P frigates (China), 120–121
export of
to Bangladesh, 92, 120, 121
to Pakistan, 30, 76, 84–85,
120–121
FC–1 fighter (China), 103, 106–107
engines for, 36, 37, 52, *58*
export of, 75–76, 81, 106–107, 136
Algerian interest in, 90
Angolan interest in, 90
Azerbaijani interest in, 91, 107
Bangladeshi interest in, 92
to Egypt, 89, 107
Sri Lankan interest in, 98, 107
to Zambia, 101
joint China–Pakistan program on,
83–84, 86, 106–107, 118
price of, 107
FL–6 missile (China), 118
FL–8 missile (China), 118
FL–9 missile (China), 118
FN–6 missile system (China), 123
foreign trading divisions, of Chinese
corporations, 2–3
4th Academy, 14
France
Chinese imports from
after arms embargo, 71, 74
campaign to end embargo, 73–74
helicopters, 67–68, *69,* 112–113
shipborne weapons systems, 68,
69
licenses for Chinese production, 43, 67,
112, 113, 114, 128
technology employed by Chinese, *69*
FT–1 satellite-guided bomb (China), 119
FT–2 satellite-guided bomb (China), 119
FT–3 satellite-guided bomb (China), 119
FT–5 satellite-guided bomb (China), 119

G

GAD. *See* General Armament
Department
GAIC. *See* Guizhou Aircraft Industry
Corporation

Galileo navigation system (European Union), *70*
GEC–Marconi Super Skyranger radar (Britain), *70*
General Armament Department (GAD), 2, 4, 5–6
General Electric, 39, 111
geopolitical period of Chinese exports (1950s–1960s), 77, 78
Germany
    Chinese imports from, *70*, 128
    technology employed by Chinese, *70*
Ghana, Chinese exports to, 93, 120
Goalkeeper 30mm artillery system (Netherlands to China), 68, 124
government control, Chinese, 3–6
Great Dragon Telecommunications Equipment Company, 17
grenade launcher (China), 135
grenades (China), 133
Grifo radar (Italy), 69, *70,* 71, 103
ground weapons
    Chinese export of, 80, 130–135
        to Bangladesh, 91
        to Bolivia, 92
        to Indonesia, 94
        to Kuwait, 80–81, 95
        to Mexico, 95
        to Pakistan, 80, 81, 86
        to Rwanda, 97
        to Saudi Arabia, 97
        to Sudan, 89
    Chinese import of, 54, *60*
        from Israel, 67
        from Russia, 54–55, *61–62*
    Chinese production of, 10–13, 130–135
    Iranian production under Chinese license, 87
Guangxi Gujiang Shipyard Company, 34
Guangzhou, shipbuilding operations in, 19, 26–27, 31–32
Guangzhou Huangpu Shipbuilding Company, 27, 32
Guangzhou Longxue Shipbuilding Company, 27, 32
Guangzhou Shipbuilding Company, 26
Guangzhou Shipyard, 31
Guangzhou Shipyard International Company, 27, 31
Guangzhou Wenchong Shipbuilding Company, 27, 32
Guizhou Aircraft Industry Corporation (GAIC), 8, 44, 116
Guizhou Liyang Aero-Engine Company (Plant No. 460), 44, 45
guns, Chinese, 134–135. *See also* small arms

H

H–6K bomber (China), 36, 44, 108
Hafei Aviation Industry Co., 8
Haixun II (Chinese patrol ship), 33
Han-class submarine (China)d, 24, *69*
Hangzhou Dongfeng Shipbuilding Company, 33
Harbin Aircraft Industry Corporation, 8, 67
Harbin Dongan Engine Manufacturing Company (DEMC, Plant No. 120), 41
heavy armor, Chinese production and export of, 124–126
Heavy Industries Taxila (Pakistan), 84
helicopter(s). *See also specific types*
    Chinese export of, 37, 111–115
        to Argentina, 91
        to Bangladesh, 92
        to Bolivia, 92
        to Cambodia, 92
        to Kenya, 94
        to Pakistan, 81, 85
        to Philippines, 97
        to Russia, 115
        to Zambia, 101
    Chinese import of
        from Europe, 67–68, *69,* 112–114
        from Russia, 52–53, *59,* 72, 112
    Chinese production of, 8–9, 10, 111–115
    engines for, 37, 41, 42–43, 111–115
helicopter carrier (China), 29
Helicopter Research & Development Institute No. 602, 8
HJ–8 anti-tank weapon (China), 85, 91, 92, 93, 96, 133

HJ–9 anti-tank weapon (China), 67, 133
HJ–73 anti-tank weapon (China), 133
howitzers. *See* artillery systems
HQ–2 missile system (China), 122
HQ–7 missile system (China), 14, 123
HQ–9 missile system (China), 14, 15, 122
HQ–12 (KS–1) missile system (China),
    14, 15, 95, 122
HQ–16 missile system (China), 122
HQ–61A missile system (China), 123
HQ–64 missile system (China), 68, 123
Huanghai Shipbuilding Corporation, 33
Huangpu Shipyard, 16, 32
Huarun Dadong Dockyard, 30–31
Huawei Technologies, 17
Hudong Heavy Machinery, 30
Hudong Shipyard, 29–31
Hudong–Zhonghua Shipbuilding Group,
    16, 26, 29–31, 85
Huifeng tactical unmanned aerial vehicle
    (China), 116
Huludao shipyards, 18, 22, 24
Hussein, Saddam, 120

I

IAI. *See* Israel Aerospace Industries
ideological period of Chinese exports
    (1950s–1960s), 77–78
Il–76MD transport (Russia to China), 52,
    56, 58, 72
Il–78 (Il–78MK) aerial refueling tankers
    (Russia to China), 52, *56, 58,* 72
imports by China, 49–75
    China as net importer, 49–51, 79
    from Europe, 65, 67–71
        aircraft engines, 36, 45, *70,* 74,
            108, 112–114
        air defense systems, 68–69,
            69–70, 71
        after arms embargo, 50, 70–71,
            74, 79
        before arms embargo, 67–69
        helicopters, 67–68, *69,* 112–114
        lifting of embargo, hope for, 71,
            73–74
        shipborne weapons systems, 68,
            69

US pressure against, 74
weapons systems relying on
    European technology,
    69–70
from Israel, 50, 65–67, 74
    cooperation on J–10 aircraft, 50,
        65, 66, 105
    ground weapons, 67
    importance to China, 67
    internal security equipment, 66,
        75
    missiles, 66, 117
    unmanned aerial vehicles, 67
    US pressure against, 50, 52,
        65–66, 105
from Russia, 51–65, 79
    aerospace, 52–53, *55–58,* 72
    aircraft engines, 36–37, 48, 52,
        57–58, 75–76, 105–109, 111
    air defense systems, 54, *60,* 122,
        123
    decline in, 50, 63, 71–73
    dollar value of, 50–51, 72–73
    good commercial terms on, 64
    ground weapons, 54–55, *61–62*
    international repercussions of, 65
    known contracts and deliveries,
        55–62
    large volume of, 63
    military–political risks of, 64–65
    no foreign components in, 64
    reliance on, 50, 70–71
    ships and naval weapons, 50,
        53–54, *58–60,* 73, 117, 118
    spending limits on, 50
    supplies in 1992–2010, 51–55
    technological restrictions on, 50,
        64–65, 73
    tight deadlines for, 64
technology transfer as priority, 74
from Ukraine
    aerial refueling tankers, 52
    aircraft carrier (unfinished), 16,
        21, 23
    aircraft engines, 84, 100, 109, 126
    missiles, 117
    naval components, 35
    Russian fighter, 106

JL–8 trainer (China). *See* K–8 trainer (China)
JL–9 trainer (China), 109–110
Jordan, Chinese exports to, 94
Joy Air Ltd., *8*

K

K–8 trainer (China), 109
  engines for, 109
  export of, 37, 109
      to Angola, 90
      to Bolivia, 92
      to Burma, 88
      to Egypt, 89
      to Ghana, 93
      to Kenya, 94
      to Morocco, 95–96
      to Namibia, 96
      to Pakistan, 80, 84
      to Sri Lanka, 98
      to Sudan, 89
      to Tanzania, 99
      to Venezuela, 101
      to Zambia, 101
      to Zimbabwe, 102
  joint China–Pakistan programs on, 84
Ka–27PS helicopter (Russia to China), 52, *59*
Ka–28 helicopter (Russia to China), 52, 53, *59*
KAB–1500 Kr guided bomb (Russia to China), 53, *57*
Kalashnikov design, 134–135
Karachi Shipyard & Engineering Works (Pakistan), 84, 120–121
KD–63 missile (China), 118–119
KD–88 missile (China), 118
Kenya, Chinese exports to, 81, 94
Kh–25 missile (Russia to China), 53, *56*
Kh–29TE missile (Russia to China), 53, *57*
Kh–31A missile (Russia to China), 53, *57*
Kh–31P missile (Russia to China), 53, *57*, 118
Kh–59ME missile (Russia to China), 53, *57*
Kh–59MK missile (Russia to China), 53, *57*

Al Khalid tank (China–Pakistan), 80, 84, 125
Khmer Rouge (Cambodia), 92
KJ–200 warning system (China), 111
KJ–2000 warning system (China), 52, 111
Komsomolsk-on-Amur Aerospace Company (Russia), 64
Korean War, 77–78
KS–1 (HQ–12) missile system (China), 14, 15, 95, 122
Kunlun (WP–14) aircraft engine, 36, 42, 46
Kunlunshan (landing platform dock), 30
Kunming shipyards, 18
Kuwait, Chinese exports to, 79, 80–81, 83, 95

L

L–7 trainer (China), 109
L–15 trainer (China), 109–110
  export of, 93, 94, 96, 102, 109–110
  JL–9 *vs.*, 109–110
  joint China–Ukraine program on, 100
Latin America. *See also specific countries*
  Chinese exports to, 75, 76, 79, 80–83
LD–2000 anti-aircraft artillery, 124
LEAP–X1C aircraft engine, 39–40
Lebanon, Chinese exports to, 107
Liaoning Shipbuilding Group, 16
licenses by China
  to Iran, 86–87
  to Pakistan, 84–85, 120–121
  to Turkey, 99–100
licenses to China
  by Britain, 44, 45
  by France, 43, 67, 112, 113, 114, 128
  by Israel, 117
  by Italy, 68, 117–118, 123
  by Russia, 80
      aircraft, 52, 63, 105
      armor, 126
      ground weapons, 54–55, 130, 132, 133, 135
  by Soviet Union
      aircraft, 102, 103, 108, 111
      armor, 124–125
      ground weapons, 130, 134
      trucks, 127
  by Swiss, 124

N

Namibia, Chinese exports to, 96, 103, 110
Nanjing Research Institute of Simulation
    Technique (NRIST), 115–116
"national champions," 4
nationalism, in Chinese defense industry,
    73
national liberation movements, 77
Navale Crotale SAM system (France to
    China), 68, 69, 86, 123
naval weapons systems. *See also specific
    weapons*
    Chinese export of, 117–119, 121
        to Bangladesh, 92
        to Burma, 88
        to Indonesia, 94
        to Pakistan, 80, 81, 85–86
    Chinese import of
        from Europe, 68, 69–70
        from Russia, 53–54, 117, 118
    Chinese production of, need for, 21, 22,
        117, 119
navy, Chinese. *See also* shipbuilding
    Chinese imports for, 17, 35
        from Europe, 68, 69–70
        from Russia, 50, 53–54, 58–60
    Chinese production for, 15–16, 18–19,
        22–33
        repair centers for, 34
        as world's largest, 22
net exporter, China as, 49, 71, 75–76, 78
Netherlands, Chinese imports from, 68,
    124
net importer, China as, 49–51, 79
NH–5 missile system (China), 123–124
NH–6 missile system (China), 123
Nigeria, Chinese exports to, 81, 96
    aircraft, 75, 96, 104, 107
    political diversification in, 83
NORICUM (Austria), 130
NORINCO Corp. *See* China North
    Industries Corporation
NORINCO Group. *See* China North
    Industries Group Corporation
Northeastern Arsenal, 41
North Korea, Chinese exports to, 82,
    89–90
    aircraft, 103

ideological period of (1950s–1960s),
    77–78
submarines, 120
Northrop Grumman, 106
North Vietnam, Chinese exports to, 77–78
Northwestern Polytechnical University
    (NWPU), 115
NRIST. *See* Nanjing Research Institute of
    Simulation Technique
nuclear submarines, Chinese production
    of, 16, 21, 24, 32
NWPU. *See* Northwestern Polytechnical
    University

O

off-road vehicles
    Chinese production and export of,
        127–129
    European technology for, 70
oil and petrochemicals, 13
Olympics (2008 Beijing), security for, 71,
    75
Oman, Chinese exports to, 96

P

P–12 ballistic missile (China), 129–130
PAC. *See* Pakistan Aeronautical Complex
Pakistan
    Bangladesh relations with, 91
    Chinese alliance against India, 83
    Chinese exports to, 49, 75, 76, 78–86
        aircraft, 75, 81, 83–86, 103, 105,
            110
        biggest recent contracts, 85
        chassis for missile system, 132
        geopolitical period of, 78
        missiles, 80, 81, 85–86, 122, 123
        ships and naval weapons, 76, 80,
            81, 84–86, 120–121
    as conduit for Western technology, 86
    joint programs with China, 83–86,
        106–107, 120–121, 126
    US sanctions against, 80
Pakistan Aeronautical Complex (PAC), 84,
    106–107
pariah states, Chinese exports to, 82

air defense systems, 54, *60*, 122, 123

  decline in, 50, 63, 71–73

  dollar value of, 50–51, 72–73

  good commercial terms on, 64

  ground weapons, 54–55, *61–62*

  helicopters, 52–53, *59*, 72, 112

  international repercussions of, 65

  known contracts and deliveries, *55–62*

  large volume of, 63

  military–political risks of, 64–65

  no foreign components in, 64

  reliance on, 50, 70–71

  ships and naval weapons, 50, 53–54, *58–60*, 73, 117, 118

  spending limits on, 50

  supplies in 1992–2010, 51–55

  technological restrictions on, 50, 64–65, 73

  tight deadlines for, 64

Chinese military–technical cooperation with, 62–65

  distinguishing features of, 63–65

  first period of (1992–1999), 62

  second period of (1999–2004), 63

  third period of (2004–), 63

Chinese shipbuilding interests of, 17, 23, 24, 29–32

licenses for Chinese production, 80

  aircraft, 52, 63, 105

  armor, 126

  ground weapons, 54–55, 130, 132, 133, 135

other defense customers of, 50, 63, 72, 73

Russian Ship Repair Works, 23

RVV–AE missile (Russia to China), 53, *56*

Rwanda, Chinese exports to, 97

S

S–300FM missile system (Russia to China), *59*

S–300PMU–1 missile (Russia to China), 54, *60*, 122

S–300PMU–2 missile (Russia to China), 54, *60*, 87, 122

SAERI. *See* Shengyang Engine Design and Research Institute

SAM (surface-to-air) missiles

  chassis for, 128–129

  Chinese export of, 122–124

    to Bangladesh, 91

    to Bolivia, 92

    to Indonesia, 94

    to Iran, 86–87, 123

    to Malaysia, 95

    to Pakistan, 80, 81, 85, 122, 123

    to Peru, 96

    to Sudan, 89

    to Turkey, 100, 122

  Chinese import of

    from Europe, 68, *69–70*

    from Russia, 53, *59, 60*

  Chinese production of, 11, 14–15, 21, 122–124

  Iranian production under Chinese license, 86–87

Sanshan New Shipyard, 33

SASAC. *See* State-owned Assets Supervision and Administration Commission

SAST. *See* Shanghai Academy of Space Flight Technology

SASTIND. *See* State Administration for Science, Technology and Industry for National Defense

satellite-guided bombs (China), 119

satellite-guided missiles (China), 129

Saudi Arabia

  Chinese exports to, 79, 81, 83, 97

  as Russia's defense customer, 63

SCAIC. *See* Sichuan Aerospace Industry Corporation

Searchwater radar (Britain), *70*, 71

self-propelled artillery (China), 131–132

Severnoye design bureau (Russia), 53

SH–1 unmanned aerial vehicle (China), 116

SH–3 unmanned aerial vehicle (China), 116

Shaanqi SX2150 truck (China), 128

Shaanxi Aircraft, 8

Shandong Huanghai Shipbuilding Company, 33

Snecma CFM International, 39, 46, 74, 111

sniper rifles, Chinese, 134–135

Song-class submarine (Germany), *70*

Soummam (Algerian training ship), 90, 121

South Asia. *See also specific countries*
Chinese exports to, 75, 76, 79, 83

Southeast Asia. *See also specific countries*
Chinese exports to, 75, 76, 79, 83

Soviet Union. *See also* Russia
breakup of, 79, 89
China's confrontation with, 78
Chinese aerospace industry aided by, 102–103
licenses for Chinese production
aircraft, 102, 103, 108, 111
armor, 124–125
ground weapons, 130, 134
trucks, 127
Sino-Pakistani alliance against, 83

spacecraft, 13–15

Space Research Corporation (Canada), 130

special-purpose aircraft, Chinese, 110–111

Sri Lanka, Chinese exports to, 49, 98
aircraft, 75, 103–104, 107
insurgency aid (1981–2009), 98

State Administration for Science, Technology and Industry for National Defense (SASTIND), 2, 4–5, 6

State Council, Chinese, 2

State-owned Assets Supervision and Administration Commission (SASAC), 2, 3–4

state-owned corporations, Chinese, 1–3.
See also specific corporations

Steyr 24M truck (China), 128

stock market, Chinese, 3

Su–27SK fighter jets (Russia to China), 52, *55*, 62, 63, 105

Su–27SK kits for assembly (Russia to China), 52, 72

Su–27UBK combat trainers (Russia to China), 52, *55*

Su–30MK2 fighters (Russia to China), 52, *59*, 63, 64, 72

Su–30MKI fighters (Russia to China), 64

Su–30MKK multirole fighters (Russia to China), 52, *54*, 63, 64

submachine gun, Chinese, 134

submarines
Chinese export of
to North Korea, 120
to Pakistan, 86, 121
to Thailand, 99
Chinese imports of
from Europe, *69, 70*
from Russia, 51, 53, *58*, 63, 73
Chinese production of, 21–22, 28
non-nuclear, 21, 24–25
nuclear, 16, 21, 24, 32

Sudan, Chinese exports to, 82, 89, 103

Sunshine unmanned aerial vehicle (China), 116

surface-to-air (SAM) missiles
chassis for, 128–129
Chinese export of, 122–124
to Bangladesh, 91
to Indonesia, 94
to Iran, 123
to Malaysia, 95
to Pakistan, 80, 81, 85, 122, 123
to Peru, 96
to Sudan, 89
to Turkey, 100, 122
Chinese import of
from Europe, 68
from Russia, 53, *59, 60*
Chinese production of, 11, 14–15, 21, 122–124
Iranian production under Chinese license, 86–87

Swiss license, for Chinese production, 124

SY–400 missile (China), 129–130

Syria, Chinese exports to, 76, 98

T

Tactical Missiles Corporation (KTRV, Russia), 53

Taihang (WS–10) aircraft engine, 36, 37, 42, 46–47, 105–106

Taishan (WS–13) aircraft engine, 37, 44–45, 107

Taiwan, Chinese naval programs and,

21–22
Taizhou Wuzhou Shipbuilding Industry
  Company, 34
Tamil Tigers (Sri Lanka), 98
tanks
  Chinese export of, 124–127
    to Bangladesh, 125
    to Burma, 88
    to Peru, 96–97
    to Sudan, 89
  Chinese–Pakistani programs on, 80, 81,
    84, 125
  Chinese production of, 11, 124–127
  weapons against, Chinese, 133
    export of, 85, 91, 92, 93, 96, 133
    Israeli technology in, 67
Tanzania, Chinese exports to, 75, 98–99, 103
Tashkent Aircraft Plant (Uzbekistan),
  52, 72
TF–8 mini-unmanned aerial vehicle
  (China), 116
TF–10 mini-unmanned aerial vehicle
  (China), 116
Thailand, Chinese exports to, 49, 80, 99
  diversification of suppliers through, 76
  ships, 30, 99, 120, 121
Tiaian Wuyue Special Vehicle, 128–129
Tiananmen Square protests, foreign
  restrictions after, 50, 79, 106
Tianjin Xingang Shipbuilding Heavy
  Industry Company, 18, 22, 26, 121
Tianyi unmanned aerial vehicle (China),
  116
Tiema XC2200 trucks (Germany), *70*
Tikhomirov Instrument Design Institute
  (Russia), 118
TL–2 missile (China), 118
TL–6 missile (China), 118
TL–10 missile (China), 118
Tongfang Jiangxin Shipbuilding Company,
  33–34
Tor–M1 missile (Russia to China), 54, *60,*
  68, 123
trainers. *See* combat trainers
transport aircraft. *See also specific types*
  Chinese import of, 52, *56, 58,* 63, 72
  Chinese production and export of, 81,
    110–111

engines for, 37, 41, 43, 110
transport helicopters. *See* helicopter(s)
trucks
  Chinese export of, 127–129
    to Angola, 90
    to Bangladesh, 91
    to Bolivia, 92
    to Burma, 88
    to Cambodia, 92
    to Democratic Republic of the
      Congo, 93
    to Ecuador, 93
    to Ghana, 93
    to Pakistan, 81
    to Peru, 96
    to Sudan, 89
    to Uganda, 100
  Chinese production of, 11, 127–129
  European technology for, *70*
Tsinghua Tongfang Corporation, 33–34
Turkey
  Chinese exports to, 83, 99–100
  production under Chinese license,
    99–100
TV–1 missile (China), 118
TY–90 missile system (China), 123

U

U8E unmanned helicopter (China), 116
UAVs. *See* unmanned aerial vehicles
Uganda, Chinese exports to, 100
Ukraine
  Chinese imports from
    aerial refueling tankers, 52
    aircraft carrier (unfinished), 16,
      21, 23
    aircraft engines, 84, 100, 109
    missiles, 117
    naval components, 35
    Russian fighter, 106
    tank engines and transmissions,
      97, 126
  interests in Chinese shipbuilding, 17
  L–15 aircraft production in, 100
Ulan–Ude aircraft plant (Russia), 112
United Arab Emirates
  Chinese exports to, 100–101

as Russia's defense customer, 63
United Kingdom
  Chinese imports from
    aircraft engines, 36, 45, *70,* 108
    air defense systems, 69, *70,* 71
    after arms embargo, 71
  licenses for Chinese production, 44, 45
  technology employed by Chinese, *70*
United States
  China as global rival of, 137
  opposition toward, and Chinese
    exports, 76
  pressure on Europe, 74
  pressure on Israel, 50, 52, 65–66, 105
  reaction to Tiananmen Square, 50, 79,
    106
  Russian exports to China and, 65
  sanctions against Pakistan, 80
unmanned aerial vehicles (UAVs)
  Chinese import from Israel, 67
  Chinese production and export of,
    115–117
UN Register of Conventional Arms, 82

V

Vanguard–1 missile system (China), 123
Vanguard–2 missile system (China),
  123–124
Varyag (aircraft carrier), 16, 21, 23
Venezuela
  Chinese exports to, 75, 76, 79, 81, 101,
    105, 110
  as Russia's defense customer, 63, 72, 73
Vietnam
  Chinese exports to (North Vietnam),
    77–78
  response to China's growing might,
    136–137
Vympel (Russia), 118

W

W–30 unmanned aerial vehicle (China),
  115–116
W–50 unmanned aerial vehicle (China),
  115–116
Wang Zhi Lin, 40

Wanshan Special Vehicle, 128
Warrior Eagle unmanned aerial vehicle
  (China), 116
weapon systems. *See specific systems*
Wen Jiabao, 111
West Germany. *See also* Germany
  Chinese imports from, *70*
WJ–5 aircraft engine, 41, 111
WJ–6 unmanned aerial vehicle (China),
  116–117
WP–7 aircraft engine, 44–45
WP–8 aircraft engine, 44
WP–14 (Kunlun) aircraft engine, 36, 42, 46
WS–2 ballistic missile (China), 129–130
WS–3 ballistic missile (China), 129–130
WS–9 (Qinling) aircraft engine, 36, 44,
  45, 108
WS–10 (Taihang) aircraft engine, 36, 37,
  42, 46–47, 105–106
WS–11 aircraft engine, 109
WS–13 (Taishan) aircraft engine, 37,
  44–45, 107
WS–18 aircraft engine, 37, 44, 111
Wuchang Shipbuilding Industry, 22,
  24–25
Wuhan shipyards, 18, 24–25
Wuhu Xinlian Shipbuilding Company, 33
WZ–5 aircraft engine, 41, 111
WZ–8A aircraft engine, 43
WZ–9 aircraft engine, 43

X

XAEC. *See* AVIC Xian Aeroengine
  (Group) Ltd.
Xian Aeroengine PLC, 38, 43–44
Xian Aircraft, 8, 111
Xianglong unmanned aerial vehicle
  (China), 116
Xian Marine Equipment Industry
  Company, 19
Xijiang Shipbuilding Company, 16, 32–33

Y

Y–7 transport (China), 37, 41, 102, 110
Y–8 transport (China)
  engines for, 37, 43, 110

export of, 110
   to Tanzania, 99
   to Venezuela, 101, 110
  production of, 102, 110–111
Y–12 transport (China)
  engines for, 37, 110
  export of, 110
   to Ghana, 93
   to Kenya, 94
   to Namibia, 96
   to Sri Lanka, 98
   to Tanzania, 99
   to Uganda, 100
   to Zambia, 101
  production of, 110
Y–20 transport (China), 110–111
Yakovlev Design Bureau, 109
Yan'an helicopters (China), 111
Yanukovich, Viktor, 100
Yilong unmanned aerial vehicle (China), 116
Yitian missile system (China), 123
YJ–91 missile (China–Russia), 118

Z

Z–2 unmanned mini-helicopter (China), 116
Z–3 unmanned mini-helicopter (China), 116
Z–5 helicopter (China), 111

Z–6 helicopter (China), 111
Z–7 helicopter (China), 111
Z–8 multirole helicopter (China), 112–113
Z–9 multirole helicopter (China), 113–114
  engines for, 37, 41, 43, 113
  export of, 37, 113–114
   to Bangladesh, 92
   to Bolivia, 92
   to Cambodia, 92
   to Kenya, 94
   to Pakistan, 81, 85
   to Philippines, 97
   to Zambia, 101
  French model for, 67, 112, 113
Z–10 attack helicopter (China), 112, 114
Z–11 multirole helicopter (China), 112, 114
Z–15 multirole helicopter (China), 112, 114–115
Zambia, Chinese exports to, 101, 107
ZDK–03 warning system (China), 111
Zhejiang Shipbuilding Corporation, 33
Zhenhua Oil, 13
Zhonghua Shipyard, 29–31
Zhongxing Telecommunications Equipment, 17
Zhuhai Chenlong Shipyard Company, 34
Zimbabwe, Chinese exports to, 101–102, 103, 107, 110
  aircraft, 75, 102, 103, 107, 110
  Western pressure and, 101–102
Zvezda–Strela (Russia), 118

About the Authors

Mikhail Barabanov graduated from the Moscow National University of Culture and then worked for the Moscow City Government. An expert on naval history and armaments. Science editor of the journal *Eksport Vooruzheniy* (Arms Exports) since May 2004. CAST researcher and Editor in Chief of the journal *Moscow Defense Brief* since 2008.

Vasiliy Kashin graduated from the Moscow State University's Institute of Asian and African Countries in 1996 and has served in various government agencies. Senior research fellow of the Russian Academy of Sciences' Institute of the Far East in 2002-2009. Deputy head of the Beijing bureau of the RIA Novosti news agency in 2010-2011. CAST researcher since 2012. Holds a PhD in Political Science.

Konstantin Makienko graduated from the Oriental Department of the Moscow State Institute of International Relations in 1995 and the French-Russian Masters' School of Political Science and International Relations in 1996. He was head of a project on conventional armaments at the Center for Policy Studies in Russia (PIR-Center) in 1996-1997. He has been Deputy Director of CAST since September 1997. He is the author of numerous articles on Russia's military-technical cooperation with other countries.

About CAST

The Centre for Analysis of Strategies and Technologies (CAST) was founded in 1997. It is a private research center specializing in the restructuring of Russia's defense industry, the national arms procurement program and the Russian arms trade. CAST is also involved in studies of the Russian army reform and armed conflicts in the former Soviet republics. CAST publishes the magazines *Eksport Vooruzheniy* (Arms Exports) in Russian and *Moscow Defense Brief* in English.

A leading Russian defense industry and arms trade think tank, CAST regularly provides consultations, analysis and other services to Russian government agencies, defense industry companies, and banks and investment firms.

About East View

The mission of East View Information Services and its imprint, East View Press, is to bring uncommon information from extraordinary places to academic, corporate, legal and government information professionals throughout the world.

East View is a leading provider of native and translated foreign language information products and services, including Russian, Chinese, and Arabic databases, print periodicals, books and microforms. The company serves all geographies and many market segments, including academic institutions, government organizations, corporations, public and federal libraries and law firms.

Readers' Notes